W9-CSP-442

Optical Properties of Glass

GLASS SCIENCE AND TECHNOLOGY

Glass Science and Technology 5

Optical Properties of Glass

IVAN FANDERLIK

State Glass Research Institute, Hradec Králové, Czechoslovakia

ELSEVIER

Amsterdam—Oxford—New York—Tokyo 1983

o 3841807

PHYSICS

Repl. TA450. F2913. 1983

Published in co-edition with
SNTL Publishers of Technical Literature, Prague

Distribution of this book is being handled by the following team of publishers

for the U.S.A. and Canada

Elsevier Science Publishing Company, Inc.
52 Vanderbilt Avenue
New York, New York 10 017

for the East European Countries, China, Northern Korea, Cuba, Vietnam and Mongolia
SNTL, Publishers of Technical Literature
Prague

for all remaining areas
Elsevier Science Publishers
1 Molenwerf

P.O.Box 211, 1000 AE Amsterdam, The Netherlands

Library of Congress Cataloging in Publication Data

Fanderlik, Ivan.
   Optical properties of glass.

   (Glass science and technology ; 5)
   Translation of: Optické vlastnosti skel.
   Bibliography: p.
   Includes index.
   1. Glass—Optical properties. I. Title. II. Series.
TA450.F2913 1983   620.1'4495   83-11644
ISBN 0-444-99652-4

IBN 0-444-99652-4 (vol. 5)
IBN 0-444-41577-7 (series)

© Ivan Fanderlik 1983
Translation © Sergej Tryml 1983

All rights reserved. No part of this publication may be reproduced, stored in a retrieval system, or transmitted in any form or by any means, electronic, mechanical, photocopying, recording or otherwise, without prior written permission of the publishers.

Printed in Czechoslovakia

CONTENTS

TA 450
F2913
1983
PHYS

5

6

The first publication dealing with the optical properties of glasses in a comprehensive way appeared in the Czechoslovak Socialist Republic in the year 1979 in a library series of glass technology manuals. Since the limited scope of that publication, and the requirement for the material to be comprehensible to less advanced readers, did not allow the author to go into more detail, the manual has now been substantially amended and considerably enlarged.

The book deals with the optical properties of glasses, viz. with the effects of the interaction of radiation with glass, with the phenomena of reflection, refraction, absorption, scatter, polarization, birefringence, and interference. These phenomena have been approached from various points of view, from the point of view of theory and measurement, and from the point of view of the dependence of optical phenomena upon chemical composition, temperature, thermal history, etc.

Attention has also been paid to those optical properties that are of decisive importance as far as commercial glass products are concerned. Chapters have been added dealing with colour generation in glasses, modification of the optical properties by the application of layers, and glasses exhibiting special optical properties (e.g., photosensitive, photoform, photochromic, polychromatic, and luminescent glasses, optical fibres, ballotini, glasses resistant to radioactive radiation, etc.).

The book will be of interest to practical glass makers, engineers and technologists in the glass industry, and scientists and designers in research institutes. It may prove helpful as a textbook for materials science students in technical colleges and undergraduates at Universities. The author hopes that the book will also be of some value to workers engaged in other branches of industry and research, viz. branches concerned with the application of the optical properties of glasses, e.g., in ocular optics, optical industry, architecture, food and pharmaceutical industries, in lighting technology, electronics, etc.

1.1 SIGNIFICANCE OF THE OPTICAL PROPERTIES OF GLASSES

The non-absorbing glasses, viz. the majority of the colourless glasses, transmit radiation in the whole range of wavelengths from the ultra-violet to the infra-red region of the spectrum. Some glasses (e.g., the chalcogenide glasses) do not transmit visible radiation but transmit infra-red radiation.

From the variation of the optical properties of glasses, conclusions can be drawn as to the influence of chemical composition, temperature or thermal history, structural arrangement, and degree of isotropy; an examination can also be made of the effect of the absorbing or emitting species, etc. Optical methods of examination have already provided us with valuable information about the structure of glasses.

Since glasses are amorphous materials, we are concerned, in contradistinction to the examination of the optical properties of the atoms or molecules, with average optical effects. As contrasted with the optical transitions of electrons in atoms, which are characterized by line spectra, and in molecules, where rotation and vibration cause the electrons to form band spectra, in solid amorphous materials, viz. glasses, the optical transitions are characterized by continuous bands invariably made up of a number of sub-bands. The effect of the surrounding fields of force makes itself felt here.

Glass is an ideal material for optical examination, and this fact has recently been taken full advantage of. For example, we might mention the important discoveries of glasses manifesting particular optical properties, e.g., photosensitive, photoform, luminescent, laser, and photochromic glasses; and glass has been used extensively as a medium for active photosensitive materials. Mention should be also made here of the discovery of glass fibres and of the wide applicability of these fibres in telecommunications technology.

The study of the optical properties of glasses is an essential prerequisite where high-quality optical systems are to be obtained; the qualitative and quantitative parametres of these systems are virtually determined by their optical properties. And the study of the optical parameters of glasses is of paramount importance for the application of glasses in lighting technology. Last but not least, optical methods find application in homogeneity tests of glasses and in studies concerned with the structure of glasses. The colour of glass is also aesthetically important, particularly from the point of view of the production of standard utility glassware, quality glass products, and imitation jewellery; suitable wavelengths of light can be filtered from both natural and artificial sources by means of specially coloured glasses.

Transparent optical coatings can greatly affect the spectral characteristics of the glass to which they are applied as far as transmission and reflection are concerned.

The author hopes that this book will provide the reader with useful introductory information for further study of the interaction of radiation with matter. The lists of references following every major chapter will enable the reader to study various aspects in depth.

1.2   SURVEY OF THE MORE IMPORTANT SYMBOLS AND UNITS

A	amplitude of light wave [m]
$\mathbf{A}$	Planckian radiator constant $2.898 \cdot 10^{-3}$ [m . K]
A'	numerical aperture of the system
$\mathbf{a}$	fractional part of interference order
a	linear absorption coefficient [m^{-1}]
a	activity of ions
a_1, a_2	dipole and quadrupole electric moments
$\mathbf{B}$	surface luminance of sphere
B	photoelastic constant [nm . cm^{-1}]
B	transmission capacity of fibre
c	velocity of light in vacuo $2.9979 \cdot 10^8$ [m . s^{-1}]
c_1	radiation constant $3.741 \cdot 10^{-8}$ [W . m^2]
c_2	radiation constant $1.439 \cdot 10^{-2}$ [m . K]
$\mathbf{c}$	concentration of absorbing centres
D	optical desity $D = -\log \tau$
D_i	internal transmission density $D_i = -\log \tau_i$
$\mathbf{D}$	diameter of scattering center
D	birefringence
d	dipole moment
$\mathbf{E}$	intensity of the electric field of light wave
E_v	illuminance [lx]
E_k, E_m	energy at state k, m
E	energy of light quantum
E_e	irradiance [W . m^{-2}]
eV	electron volt
e	radiant exposure [W . s . m^{-2}]
f_i	power of oscillator
f	focal distance
ΔG	change in free enthalpy in the transition of liquid to crystal at T
H	light exposure [lx . s]
H	change in enthalphy

h	Planckian constant $6.6256 \cdot 10^{-34}$ [J . s]
$\mathbf{h}, \mathbf{h_0}$	rate of cooling
I	number of stable nuclei formed in a specific volume per unit of time
I_v	luminous intensity [cd]
I_e	radiant intensity [W . sr^{-1}]
k	wave number
$\mathbf{k}$	Boltzmann constant $1.380 \cdot 10^{-23}$ [J . K^{-1}]
k'	absorption index
k_1	equilibrium constant
$\mathbf{l}$	angular momentum quantum number
l	thickness
L_v	luminance [nt]
L_e	radiance [W . m^{-2} . sr^{-1}]
M	molar mass
$\dfrac{M}{\varrho}$	molar volume (M_v)
$M_{e,\lambda}$	monochromatic emission
$\mathbf{m}$	magnetic quantum number
m	mass of particle
m'	number of boundary surfaces
m	gradient
$\mathbf{n}$	principal quantum number
$\mathbf{n_r}$	radial quantum number
$\mathbf{n_\varphi}$	azimuthal quantum number
n	integer multiple of light quantum
N	mole
n	refractive index of glass
n'	refractive index of medium
n''	number of atoms in specific volume
$n(\lambda)$	refractive index related to wavelength λ
$n_F - n_C, n_{F'} - n_{C'}$	mean dispersion
$N_\mathbf{A}$	Avogadro number $6.022 \cdot 10^{26}$ [kmol^{-1}]
Δn	difference between indices of refraction (change in refractive index)
$\dfrac{\Delta n}{\Delta \lambda}$	characteristic dispersion of prism
$\mathbf{N}$	number of scattering centres in specific volume
P	probability
p_e	excitation purity
p_c	colorimetric purity
p_{H_2O}	water steam pressure
p_1	dipole magnetic moment
Q_v	quantity of light [lm . s]

R	reflection factor
R	specific refraction
$\mathbf{R_M}$	molar refraction
$\mathbf{R}_i$	ionic refraction
R	radius of scattering centre
Re 90°	Rayleigh ratio
s	spin quantum number
s, p, d, f	electronic configuration in Paschen's notation
ΔS	change in entropy
t	time
T_t	melting temperature
T	absolute temperature $[K]$
U	potential energy
u, v	coordinates of colorimetric triangle of equal chrominances
v	velocity of particles
V	volume
V_m	volume of gram atom of crystal
v_λ	phase velocity of ray of light
$V_{(\lambda)}$	relative luminous efficiency of the human eye
V_g	group velocity
$W*$	thermodynamic nucleation barrier
x, y, z	coordinates
X, Y, Z	reference coordinates
$\bar{x}(\lambda), \bar{y}(\lambda), \bar{z}(\lambda)$	spectral tristimulus values
x, y, z	chromaticity coordinates
$\ddot{x}$	acceleration
z/a^2	field intensity
z	dissymmetry
α	absorption factor (absorptance)
α_i	internal absorptance
α_0	coefficient of static polarizability
α	coefficient of optical polarizability
$\dfrac{\Delta\alpha}{\Delta\lambda}$	dispersion of grating
β	propagation constant
β	polarizability tensor
δ_m	minimum deviation between incident and emerging ray
$\dfrac{\Delta\delta}{\Delta\lambda}$	angular dispersion of prism
δ	phase difference
$\varDelta$	path difference
ε	molar linear absorption

13

ε_0 static permittivity

$$\nabla^2 = \frac{\partial^2}{\partial x^2} + \frac{\partial^2}{\partial y^2} + \frac{\partial^2}{\partial z^2}$$ Laplacian operator

η	quantum yield
ϑ'	optical path
ϑ	actual length of path
λ	wavelength
λ_{max}	wavelength maximum radiant intensity
ν	frequency
ν_0	vibration frequency of electrons
ν_e, ν_d	Abbe value
$\Delta\nu$	width of line of spontaneous emission
ϱ	specific mass (density)
ϱ	reflectance
ϱ_r	direct reflectance
ϱ_d	scattered reflectance
ϱ'	depolarization
$\sigma.\ \pi,\ \delta,\ \varphi$	electron orbitals
σ	Stefan-Boltzmann full radiation constant $5.67 . 10^{-8}\ [\mathrm{W} . \mathrm{m}^{-2} . \mathrm{K}^{-4}]$
σ'	interphase free energy at the boundary between crystal and liquid
δ	stress in glass
τ	transmission factor (transmittance)
τ_r	direct transmittance
τ_d	scattered transmittance
τ	fictive temperature
τ_i	internal transmission factor
τ'	decay time of luminescence
φ_1, φ_2	phase shift
χ	coefficient of isothermal compressibility
ψ	scalar quantity — wave function
ψ_m, ψ_k	wave function at energy state m, k
Φ	incident luminous flux
Φ_e	radiant flux $[\mathrm{W}]$
$\Phi_{\varrho r}$	mirror reflected luminous flux
$\Phi_{\varrho d}$	scattered reflected luminous flux
$\Phi_{\tau r}$	regularly traversing luminous flux
$\Phi_{\tau d}$	scattered traversing luminous flux
Φ_α	luminous flux absorbed
Φ	function
Φ_p	threshold intensity
ω	angular frequency
ω	refracting angle of prism

Θ_1	angle of incidence of luminous flux
Θ_1'	angle of reflection of luminous flux
Θ_2	angle of refraction of luminous flux
Θ_m	critical angle
Θ_p	Brewster polarizing angle

2.1   INTRODUCTION

For an understanding of the optical properties of glasses, particularly those of a more theoretical nature, it is desirable that the reader should have at least some basic knowledge of the theory of the quantum mechanics of the atom. Chapter 2 presents a brief survey of the principles of quantum mechanics. For more detailed information about that subject the reader is referred to the references at the end of this chapter.

2.2   BOHR'S MODEL THEORY OF THE ATOM AS MODIFIED BY SOMMERFELD

Bohr's theory was modified by Sommerfeld, who substituted two conditions concerning elliptic electron orbits and introduced radial n_r, and azimuthal n_φ quantum numbers for those ellipses [1, 2, 3, 10]. The energy, angular momentum, and spin of these electrons are determined by four quantum numbers:

n – principal quantum number ($n = n_r + n_\varphi$),

l – angular momentum quantum number,

m – magnetic quantum number,

s – spin quantum number.

The principal quantum number, **n**, gives the energy of an electron, and the angular momentum quantum number, **l**, the shape of the path (orbit) of an electron; for given **n**'s the latter can assume values ranging from zero to **n** − 1.

The magnetic quantum number, **m**, which for a given 1 can assume values −**1** ... 0 ... +**1**, defines all the possible position of the orbit of an electron. The spin quantum number, **s**, gives the spin of an electron, its values being ±**1/2**.

But the applicability of the Bohr−Sommerfeld theory was limited solely to the atomic spectra of hydrogen and elements similar to hydrogen, viz. those having only one valency electron. It failed to apply to atoms containing a larger number of electrons.

2.3 DE BROGLIE'S THEORY AND THE SCHRÖDINGER EQUATION

De Broglie's theory [1, 2, 3, 10] initiated a completely new approach, associating the motion of each mass particle with the propagation of a wave of length

$$\lambda = \frac{h}{mv},$$ (1)

where h is the Planckian constant $(6.6256 \pm 0.0005) \cdot 10^{-34}$ [J . s],

m = the mass of the particle,

and v = the velocity of the particle.

De Broglie's theory was applied by Schrödinger, who derived an equation with the help of which the wave function of electrons in atoms can be determined. Starting from the wave equation

$$\mathbf{E} = A \cos 2\pi(vt - k\,\mathbf{x}),$$ (2)

where $\mathbf{E}$ is the intensity of the wave field,

A is the amplitude of the light wave,

v is the frequency,

t is the time,

k is the wave number,

and $\mathbf{x}$ is the coordinate in the direction of the wave propagation.

Differentiating with respect to time t and coordinate $\mathbf{x}$, we can write

$$\frac{\partial^2 \mathbf{E}}{\partial t^2} = -4\pi^2 v^2 A \cos 2\pi(vt - k\mathbf{x}) = -4\pi^2 v^2 \mathbf{E},$$ (3)

$$\frac{\partial^2 \mathbf{E}}{\partial \mathbf{x}^2} = -4\pi^2 k^2 A \cos 2\pi(vt - k\mathbf{x}) = -4\pi^2 k^2 \mathbf{E}.$$ (4)

Thus

$$\frac{\partial^2 \mathbf{E}}{\partial t^2} = \frac{v^2}{k^2} \frac{\partial^2 \mathbf{E}}{\partial \mathbf{x}^2}.$$ (5)

If $k = \dfrac{1}{\lambda} = \dfrac{v}{c}$, then

$$\frac{\partial^2 \mathbf{E}}{\partial t^2} = c^2 \frac{\partial^2 \mathbf{E}}{\partial \mathbf{x}^2}.$$ (6)

Equation (6) is the wave equation for a plane wave propagating in the direction $\mathbf{x}$ at velocity c. Determining $\dfrac{\partial^2 \mathbf{E}}{\partial \mathbf{x}^2}$, we get

$$\frac{\partial^2 \mathbf{E}}{\partial \mathbf{x}^2} = -4\pi^2 k^2 \mathbf{E} = -\frac{4\pi^2}{\lambda^2} \mathbf{E}.$$ (7)

In the general case of a spherical wave, in which some scalar quantity ψ changes periodically, the motion of the wave can be resolved into the directions **x, y, z**. The wave equation then assumes the following forms:

$$\frac{\partial^2 \psi}{\partial t^2} = c^2 \left(\frac{\partial^2 \psi}{\partial x^2} + \frac{\partial^2 \psi}{\partial y^2} + \frac{\partial^2 \psi}{\partial z^2} \right), \tag{8}$$

or

$$\frac{\partial^2 \psi}{\partial x^2} + \frac{\partial^2 \psi}{\partial y^2} + \frac{\partial^2 \psi}{\partial z^2} + \frac{\pi^2}{\lambda^2} \psi = 0. \tag{9}$$

The sum of three second derivatives with respect to three coordinates is denoted as ∇^2; thus

$$\frac{\partial^2 \psi}{\partial x^2} + \frac{\partial^2 \psi}{\partial y^2} + \frac{\partial^2 \psi}{\partial z^2} = \nabla^2 \psi, \tag{10}$$

where ∇^2 is the Laplacian operator.
The form of the wave equation is then

$$\nabla^2 \psi + \frac{4\pi^2}{v^2} \psi = 0. \tag{11}$$

Substituting the value for the wavelength from de Broglie's equation (1), expressing velocity v with the help of the energy of the electrons, and the total energy of the electron, **E**, as

$$\mathbf{E} = \frac{mv^2}{2} + \quad , \tag{12}$$

where U stands for potential energy,
and using for λ the relationship

$$\lambda = \frac{h}{\sqrt{2m(\mathbf{E} - U)}}, \tag{13}$$

then substituting from equation (13) into equation (11), we arrive at *Schrödinger's equation*

$$\nabla^2 \psi + \frac{8\pi^2 m}{h^2} (\mathbf{E} - U) \psi = 0. \tag{14}$$

For a free electron, the potential energy $U = 0$. If the electron moves in direction, **x**, then

$$\frac{\partial^2 \psi}{\partial x^2} + \frac{8\pi^2 m}{h^2} \mathbf{E}\psi = 0, \tag{15}$$

and solving (15) we get

$$\psi = \psi_0 \cos \frac{2\pi}{h} \sqrt{(2m\mathbf{E})}\, \mathbf{x}. \tag{16}$$

The energy **E** can assume any value. The dependence of the function ψ upon time is included in ψ_0.

18

When the potential energy, U, is different from zero, the Schrödinger equation will prove satisfactory only for certain total energies of the electron, **E**, viz. the electron can assume only certain discrete quantum values of energy.

Potential well and harmonic oscillator

In simple cases, the characteristic values of the energy can be computed directly, without the help of the Schrödinger equation.

An electron the kinetic energy of which is in the range of 0 to a' placed in a so-called *potential well* cannot leave this region of potential energy, for outside the well the potential energy of that electron would be infinite. The probability of finding an electron outside the potential well will thus be equal to zero.

In the potential well the electron waves impinging upon the walls of the well rebound to form a system of standing waves for which we can write

$$\psi = A \sin 2\pi \frac{x}{\lambda} + B \cos 2\pi \frac{x}{\lambda}, \tag{17}$$

where λ is the wavelength, and the dependence of ψ upon time is included in A and B. Since at the boundaries of region 0 to a', ψ equals zero, which indicates that at points $x = 0$ and $x = a'$ there are nodes, we can write for $x = 0$

$$\varphi = B = 0, \tag{18}$$

and

$$\psi = A \sin \frac{2\pi x}{\lambda}. \tag{19}$$

For $x = a'$ then

$$\psi = A \sin \frac{2\pi a'}{\lambda} = 0, \tag{20}$$

so that

$$\frac{2\pi a'}{\lambda} = n\pi, \tag{21}$$

where n is an arbitrary integer.

The length of the standing waves can be

$$\lambda = 2a', a', \frac{2a'}{3}, \frac{a'}{2}, ..., \frac{2a'}{n}. \tag{22}$$

The *velocity of an electron* in a potential well is given by the equation

$$v = \frac{h}{m\lambda} = \frac{hn}{2ma'}, \tag{23}$$

where n equals 1, 2, 3, ...,

and h is the Planck's constant;

the *energy of the electron* is defined by the equation

$$E = \frac{mv^2}{2} = \frac{n^2h^2}{8ma'^2},$$ (24)

where m is the mass of the electron,
and v is the velocity of the electron.

Thus, the energy and the velocity of an electron are quantized and can assume values given by equations (23) and (24) only. n is the quantum number. The energy levels $n = 1, 2, 3$, and the probability ψ^2 of finding an electron at a certain place in the potential well under given conditions are shown in Figure 1.

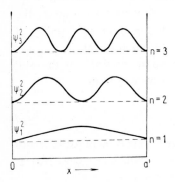

Fig. 1 Potential wells.   Fig. 2 Harmonic oscillator.

Now, let us assume that a particle (electron) is exposed to the effect of an external force (elasticity), which makes the particles (electrons) return to equilibrium position $\mathbf{x} = 0$, and the value of which is proportional to the deviation of that particle from that equilibrium position (see Fig. 2).
We can then write

$$f = -k\mathbf{x},$$ (25)

where k is the force constant (of elasticity),
and f is the elastic force.
The particle displaced by that force then *oscillates harmonically.* If

$$m\frac{d^2\mathbf{x}}{dt^2} = -k\mathbf{x},$$ (26)

then

$$x = x_0 \cos 2\pi vt,$$ (27)

because the *velocity* of the movement of the particle (first derivative with respect to time) equals

$$v = \frac{d\mathbf{x}}{dt} = -2\pi v\mathbf{x}_0 \sin 2\pi vt,$$ (28)

20

and the *acceleration* (second derivative)

$$\frac{d^2\mathbf{x}}{dt^2} = -4\pi^2 v^2 \mathbf{x}_0 \cos 2\pi vt. \tag{29}$$

Substituting from equations (27) and (29) into equation (26) and reducing, we get

$$4\pi^2 v^2 m = k. \tag{30}$$

If the frequency of the vibration is

$$v = \frac{1}{2\pi} \sqrt{\frac{k}{m}}, \tag{31}$$

then the higher the coefficient k, the more violent will be the vibration, and a larger mass of the vibrating particle m will in turn lead to more restrained vibration.

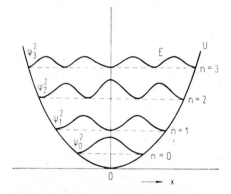

Fig. 3 Potential and total energies of an oscillator and squared characteristic functions.

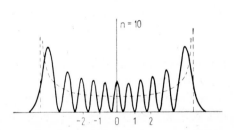

Fig. 4 Quantum mechanical (—) and classical (- - -) probabilities of the position of an oscillator for $n = 10$.

The potential energy of the particle subjected to an elastic force $f = -k\mathbf{x}$ equals

$$U = \frac{k\mathbf{x}^2}{2}, \tag{32}$$

and such a particle is termed a *classical harmonic oscillator*, whose energy

$$E = \frac{k\mathbf{x}^2}{2} + \frac{mv^2}{2}. \tag{33}$$

For a *quantum oscillator*, however, we get a slightly different result. If we define the dependence of potential energy U upon displacement $\mathbf{x}$, as we did in the case of the potential well mentioned above, equation (32) will give us a parabola (see Fig. 3).

In this well, the standing waves will assume shapes which can be ascertained only with the help of the Schrödinger equation; the possible values of the total

21

energy of the harmonic oscillator, quantized again, are hereby determined as well. Total energy

$$E_n = \left(n + \frac{1}{2}\right) h\nu, \tag{34}$$

where the values of the vibration quantum number are $n = 0, 1, 2, 3 \ldots$. Figure 3 shows the levels of total energy according to equation (34) and the probability of the respective states of the oscillators. If this probability is compared with the probability of finding the classical oscillator (Fig. 4), then with higher quantities of energy the approximation of the two curves of probability (the classical and the quantum) will be proportional to the magnitude of quantum number n. This is the so-called *principle of correspondence*, formulated by Bohr, stating that with high quantum numbers (large energies) the laws of quantum mechanics will coincide with those of classical mechanics. On the other hand, with smaller quantum numbers the difference between the two interpretations will be particularly marked.

Vibration never ceases and the probability of finding an oscillator at points other than those of the equilibrium position $x = 0$ is finite even at absolute zero temperature. The so-called zero point energy is thus different from zero. Classical theory, on the contrary, claims certainty for the position of a stationary pendulum at point $x = 0$, and zero probability at other points.

The existence of zero-point vibration energy has been proved by experiments for any systems capable of a vibrating motion (molecules, crystals).

As already mentioned, the solution of the Schrödinger equation applies for certain privileged values of the energy of the electron only, yielding the quantum numbers n, l, m. It does not, however, account for the existence of the fourth degree of freedom of the electrons, which is given by the intrinsic angular momentum arising from spin. Dirac was the first to generalize this equation using relativistic quantum mechanics.

For a system of more than two electrons the Schrödinger equation must be solved approximatively. This approach takes the electrons of a multi-electron system to be independent of one another; further approximation will then take account of the mutual interactions, but the assumption is that the state of the system will not undergo big changes at zero approximation. The method of successive approximation was applied, e.g. in the interpretation of the atomic spectra. The squared absolute value of the wave function thus expresses the probability of an electron being found at a certain place. For the probability P stating that an electron will be found in volume dV surrounding the point x, y, z, we can write

$$dP = |\psi(x, y, z)^2| \, dV. \tag{35}$$

Figure 5 gives polar diagrams of the density of the probability of the occurrence of electrons s, p, d, f [2, 3, 10]. They are virtually sections of rotating bodies.

From Figure 5 it follows that the s electrons manifest spherical symmetry and that all directions of occurrence are thus equally probable. With electrons $p, d, f,$

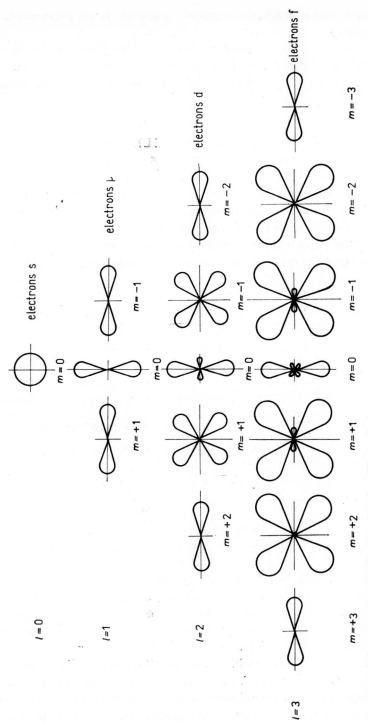

Fig. 5 Polar diagrams of probability densities for different l's.

there are some more probable directions of occurrence, which are of some importance in the bonding and formation of molecules.

The orbitals with quantum number $l = 0$ are in this case denoted by s, those with quantum number $l = 1$ by p, with quantum number $l = 2$ by d, and with quantum number $l = 3$ by f. And, similarly, for $m = 0, 1, 2, 3$, the orbitals are denoted by σ, π, δ, φ, respectively.

2.4 CONFIGURATION OF THE ELECTRONS IN AN ATOM

In an atom the configuration of the electrons in the different orbits is based upon the classification of the elements. The number that an element has in the periodic system coincides with the number of the electrons in its atom. This number is a most important aspect of the atomic structure, for it is upon the lay-out of the electron shells that the physico-chemical properties of an atom depend.

Table 1

	1s	2s	2p			3s	3p			
Na	↑↓	↑↓	↑↓	↑↓	↑↓	↑				$1s^2\,2s^2\,2p^6\,3s$
Mg	↑↓	↑↓	↑↓	↑↓	↑↓	↑↓				$1s^2\,2s^2\,2p^6\,3s^2$
Al	↑↓	↑↓	↑↓	↑↓	↑↓	↑↓	↑			$1s^2\,2s^2\,2p^6\,3s^2\,3p$
Si	↑↓	↑↓	↑↓	↑↓	↑↓	↑↓	↑	↑		$1s^2\,2p^2\,2p^6\,3s^2\,3p^2$
P	↑↓	↑↓	↑↓	↑↓	↑↓	↑↓	↑	↑	↑	$1s^2\,2s^2\,2p^6\,3s^2\,3p^3$
S	↑↓	↑↓	↑↓	↑↓	↑↓	↑↓	↑↓	↑	↑	$1s^2\,2s^2\,2p^6\,3s^2\,3p^4$
Cl	↑↓	↑↓	↑↓	↑↓	↑↓	↑↓	↑↓	↑↓	↑	$1s^2\,2s^2\,2p^6\,3s^2\,3p^5$
Ar	↑↓	↑↓	↑↓	↑↓	↑↓	↑↓	↑↓	↑↓	↑↓	$1s^2\,2s^2\,2p^6\,3s^2\,3p^6$

The periodicity of the properties of the elements is mirrored by the periodicity of the lay-cut of the electron shells, so that atoms grouped together in the periodic system may differ in the number of their electrons, but their electron shells will manifest full similarity. Maximum stability, viz minimum energy, of the different configurations can be achieved so that the orbits are gradually filled in the order of increasing energy. But Pauli's exclusion principle states that in each orbital there can only be two electrons with antiparallel spins. If electrons appear in degenerate orbits (orbits of equal n have equal energy), they do not get paired in single orbits until all the orbits of that level have been occupied by odd electrons with parallel spins (Hund's rule). With the help of these rules we can determine the configuration of the electrons of an element in its ground state. As already mentioned above, two electrons in an atom must differ in at least one of

24

the quantum numbers **n, l, m, s**. Thus, if for two electrons the quantum numbers **n, l, m** are equal, the difference must lie with the spins (as shown in Table 1).

In view of the fact that at first sight the notation of the number of the electrons and their lay-cut in the atoms may seem a bit complicated, Table 2 introduces Paschen's notation (giving the distribution of the electrons into the different energy levels), and Russel-Saunders'notation (describing the ground state of the atoms and their excited states) for several selected atoms.

Table 2

Atomic number	Element	Electronic configuration in Paschen's notation	Total spin quantum number of atom S	Multiplicity of terms $\varkappa$	Total orbital quantum number of atom L		Total internal quantum number of atom J	Ground term of atom in Russel-Saunders notation
1	H	1s	1/2	2	0	S	1/2	$^2S_{1/2}$
2	He	1s²	0	1	0	S	0	$^1S_0$
3	Li	1s² 2s	1/2	2	0	S	1/2	$^2S_{1/2}$
4	Be	1s² 2s²	0	1	0	S	0	$^1S_0$
5	B	1s² 2s² 2p	1/2	2	1	P	1/2	$^2P_{1/2}$
6	C	1s² 2s² 2p²	1	3	1	P	0	$^3P_0$
7	N	1s² 2s² 2p³	3/2	4	0	S	3/2	$^4S_{3/2}$
8	O	1s² 2s² 2p⁴	1	3	1	P	2	$^3P_2$
9	F	1s² 2s² 2p⁵	1/2	2	1	P	3/2	$^2P_{3/2}$
10	Ne	1s² 2s² 2p⁶	0	1	0	S	0	$^1S_0$
21	Sc	1s² 2s² 2p⁶ 3s² 3p⁶ 3d 4s²	1/2	2	2	D	3/2	$^2D_{3/2}$
22	Ti	1s² 2s² 2p⁶ 3s² 3p⁶ 2d² 4s²	1	3	3	F	2	$^3F_2$
23	V	1s² 2s² 2p⁶ 3s² 3p⁶ 3d³ 4s²	3/2	4	3	F	3/2	$^4F_{3/2}$
24	Cr	1s² 2s² 2p⁶ 3s² 3p⁶ 3d⁵ 4s	3	7	0	S	3	$^7S_3$
25	Mn	1s² 2s² 2p⁶ 3s² 3p⁶ 3d⁵ 4s²	5/2	6	0	S	5/2	$^6S_{5/2}$
26	Fe	1s² 2s² 2p⁶ 3s² 3p⁶ 3d⁶ 4s²	2	5	2	D	4	$^5D_4$
27	Co	1s² 2s² 2p⁶ 3s² 3p⁶ 3d⁷ 4s²	3/2	4	3	F	9/2	$^4F_{9/2}$
28	Ni	1s² 2s² 2p⁶ 3s² 3p⁶ 3d⁸ 4s²	1	3	3	F	4	$^3F_4$
29	Cu	1s² 2s² 2p⁶ 3s² 3p⁶ 3d¹⁰ 4s	1/2	2	0	S	1/2	$^2S_{1/2}$
30	Zn	1s² 2s² 2p⁶ 3s² 3p⁶ 3d¹⁰ 4s²	0	1	0	S	0	$^1S_0$

2.5   ELECTRON TRANSITIONS

The state of a quantum system is thus defined by the wave function and by the energy of the system appurtenant to it in a given state. We will, quite understandably, be most interested in the wave functions of the valency electrons, viz.

electrons of the outer shells surrounding the nucleus. The energy of these electrons is within the optical range, viz. within the range of $1-4$ eV. The other electrons are bonded to the nucleus by much larger forces, so that the changes in the states of these electrons give rise to the emission of shorter-wave radiation, which is more energy-intensive.

If a quantum system (atom, molecule, crystalline or amorphous material — glass) is impinged upon by an electromagnetic wave the electric intensity of which is moderate as compared with the inner fields of the system, and the wavelength of which is considerably longer that the linear dimensions of the system [10], the system will develop a bipolar electric moment

$$d = \varepsilon_0 \beta \mathbf{E}, \tag{36}$$

which is proportional to the intensity of the electric field $\mathbf{E}$. The coefficient of proportionality, β, is a symmetric tensor of second order, called the polarizability tensor. ε_0 denotes permittivity of free space.

The so-called *oscillator strength* characterizing the quantum transitions in the system is positive, as far as the transitions from the ground state are concerned. The latter determine the polarizability of the system in its ground state. The theoretical calculation of the oscillator strengths requires the knowledge of the respective functions of the states between which the transitions take place. For the simpler quantum systems these functions are known, for the more complex systems the wave functions must again be determined by approximate methods.

The states of the electrons in an atom thus correspond to the outer energy conditions that influence the atom. With the energy conditions varying, the electrons pass from one state into another.

The majority of amorphous materials obtained from the melt by supercooling are insulators or, in some cases, semi-conductors, with a broad forbidden band of energies, the width of which exceeds 1 eV. In Figure 6 the reader will find schematic representations [9, 7] of energy states of optical electrons undergoing transitions in an atom (a), molecule (b), crystal (c), and amorphous material (d) [18].

Thus, if the system changes from the m-th to the k-th state under the influence of external energy, viz. wave function ψ_m changes to ψ_k, and the energy of the system E_m to E_k, the energy gap will be balanced, according to the law of the conservation of energy, either by the absorption of the given amount of energy (of the photon of energy $E_k - E_m > 0$), or, vice versa, by the release (emission) of the photon of energy $E_m - E_k$. For this energy we can write

$$E_m - E_k = h\nu. \tag{37}$$

It is, however, important to note that the energy transitions of the electrons in amorphous materials, viz. glasses, are more or less influenced by the field of force of the surrounding ions (atoms). But these problems will be discussed in more detail below. All the energy levels are not equally eligible for transitions to take

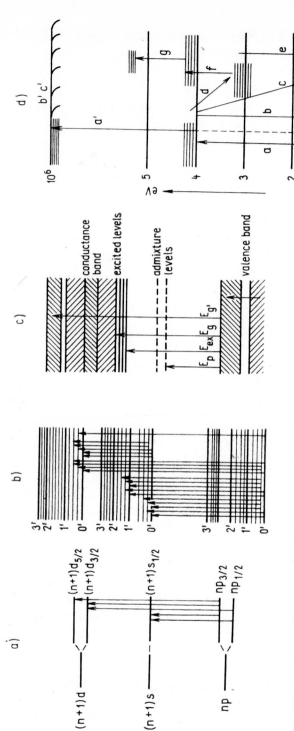

Fig. 6 Energy states of the optical electrons undergoing transitions in an atom (a); molecule (b); crystal (c); and amorphous material (d), where: a — absorption in the ultra-violet region of the spectrum, energy exceeding 4 eV; b — the excited electrons return from state a into the ground state spontaneously or prompted by external effects. It is a spontaneous emission invariably in the visible region of the spectrum or, c — the transition from level a into the ground state is a non-radiative transition (conversion to thermal energy); d — owing to the transition from level a to the admixture level of the colour centres, the energy of the radiated photon is considerably lower than the energy of the photon absorbed; e — the transition from the admixture level of the colour centres has its effects in the visible region of the spectrum. Transitions can also occur after a lapse of time (phosphorescence).

Stimulated emission takes a similar course. f, g — the presence of the admixture levels of the colour centres gives rise to transitions in the visible region of the spectrum; a', b', c' — the $\sim 10^6$ eV energies (e.g., β, γ, X-ray radiation) being absorbed, the process of the excitation of the electrons (a') into b', c' states and the formation of the metastable colour centres are in progress. Spontaneous transition to the initial state level is extremely slow but may be accelerated by temperature.

27

place between any of them. If the transition between two given energy levels is to be finitely probable, certain conditions, called selection rules, will have to be satisfied (as already mentioned above).

The energy released during transition need not however, in all cases be *radiative*. It can undergo conversion into thermal energy, and the transition is then termed *non-radiative*, and vice versa. Transitions that do not satisfy the selection rules are called *forbidden* transitions; in quantum systems the probability of their occurrence is very low.

Absorption, viz. a transition from a lower into a higher energy state, always requires some energy to fill the energy gap.

For the *emission of light* two qualitatively different modes of transition must be considered:

(a) spontaneous transitions,

(b) stimulated (forced) transitions.

In Figure 7 all the three modes of the transition of electrons are schematically represented.

Under absorption, level k is occupied in the initial state. Under the effect of energy $h\nu_{km}$ level m is filled after transition, and the energy absorbed. In colourless glasses, absorption affects the ultraviolet and infra-red regions of the spectrum, while in coloured glasses the visible region of the spectrum also depends upon the absorptive capacity and concentration of the colouring substances introduced.

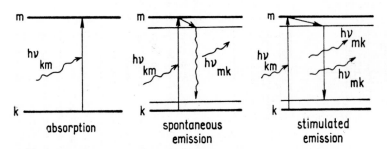

Fig. 7 Three basic types of quantum transitions between two levels.

For spontaneous emission, m is the initial state of the system. The electrons have been brought into this state by pumping; they immediately pass over into a lower metastable level by a non-radiative transition. The occupation of level k and a newly formed photon constitute the final state. Luminescence in glasses is an example of spontaneous emission.

For stimulated emision, the system is initially in state m. Here, too, the electrons have been brought into this state by pumping. They have, however, immediately passed over into a lower metastable level by non-radiative transitions. Pumping is effected with the help of flash tubes, continuous discharge tubes, or special lamps. Together with the photon that has not been absorbed two other photons leave the system. Stimulated emission is conditional upon inversion, viz. the higher energy level must manifest a higher concentration of electrons than the lower level.

Another difference between spontaneous and stimulated emissions is the fact that spontaneous emission is non-coherent and noise-like in its character. The importance of coherence will be explained in Section 4.5.6. In forced emission, on the other hand, the frequency corresponds to the energy difference between the levels, and the physical properties of forced radiation are preserved.

References

[1] M. V. Volkenštejn, Struktura a fyzikální vlastnosti molekul (Structure and Physical Properties of Molecules). NČSAV, Prague 1962.

[2] M. G. Veselov, Úvod do kvantové teorie atomů a molekul (Introduction to Quantum Theory of Atoms and Molecules). SNTL, Prague 1966.

[3] V. Trkal, Stavba atomů a molekul (Structure of Atoms and Molecules). SNTL, Prague 1968.

[4] M. Garbury, Optical Physics. Academic Press, New York—London 1965.

[5] M. V. Volkenstejn, Molekularnaya optika (Molecular Optics). GITTL, Moscow 1951.

[6] A Sommerfeld, Optics. Academic Press, New York 1954

[7] A. R. Hippel, Molekulová fyzika hmoty (Molecular Physics of Matter). SNTL, Prague 1963.

[8] V. Šatava, Úvod do fyzikální chemie silikátů (Introduction to Physical Chemistry of Silicates). SNTL, Prague 1965.

[9] V. Prosser, Optické vlastnosti pevných látek (Optical Properties of Solids). SPN, Prague 1971.

[10] A. S. Davydov, Kvantová mechanika (Quantum Mechanics). SPN, Prague 1978.

[11] E. U. Condon and H. G. Shortley, The Theory of Atomic Spectra. Cambridge University Press, London—New York 1951.

[12] P. A. M. Dirac, The Principles of Quantum Mechanics. Clarendon Press, Oxford 1958.

[13] M. Jammer, The Conceptual Development of Quantum Mechanics. McGraw-Hill, New York 1966.

[14] F. Hund, Geschichte der Quantentheorie. Wissenschaftsverlag, Wien—Zürich 1975.

[15] J. R. Klauder and E.C.G. Sudarshan, Fundamentals of Quantum Optics. W. A. Benjamin, New York—Amsterdam 1968.

[16] L. D. Landau and Je. M. Lifshic, Kvantovaya mekhanika (Quantum Mechanics). Nauka, Moscow 1972.

[17] B. L. Van der Waerden, Source of Quantum Mechanics. North-Holland Publ. Co., Amsterdam 1967.

[18] W. Eithel, Silicate Science, Vol. 2. Academic Press, New York—London 1965.

[19] D. R. Hartree, The Calculation of Atomic Structures. Wiley, New York 1957.

[20] D. Böhm, Quantum Theory. Prentice-Hall, New York 1952.

3   RADIATION

3.1   INTRODUCTION

The present chapter on radiation follows Chapter 2, which dealt with the quantum mechanics of the atom and the transitions of the electrons. The transitions of electrons are initiated by radiant energy, and the qualitative and quantitative changes in radiant energy are characteristic of an active medium — glass, whether we consider radiation from the point of view of quantum or wave mechanics.

3.2   DEFINITION OF RADIATION

Radiation can in general be defined as the propagation of energy through space. From the point of view of the electromagnetic theory, radiation is defined by wavelength λ, or by the number of oscillations per unit of time, viz. by frequency v. The velocity of the propagation is equal for all modes of radiation, and in vacuo it equals

$$2.99792 \pm 0.00003 . 10^8 \, [\mathrm{m . s^{-1}}].$$

For the relationship between wavelength λ and frequency v, we have

$$\lambda = \frac{c}{v}, \tag{38}$$

where c denotes the velocity of light in vacuo.

In the physics of electromagnetic waves quantities are often used which give the number of the waves per unit of length in the direction of their propagation, viz. wave number k,

$$k = \frac{1}{\lambda}. \tag{39}$$

Figure 8 gives the whole radiation spectrum arranged in order of the wavelength.

The range of *visible radiation* varies with the observer. It is generally defined as lying within the $380-400$ nm and $760-780$ nm wavebands.

Ultra-violet radiation is defined by wavelengths that are shorter than those of the visible region and longer than 1 nm. On the other hand, the *infra-red region* is limited by wavelengths that are longer than the wavelengths of the visible region of the spectrum but shorter than 1 mm.

We are concerned with visible radiation. From the physical point of view, luminous radiation is the visible part of so-called optical radiation. Although all parts of electromagnetic spectra are basically the same physically, the different modes of radiation materialize in ways that are completely different.

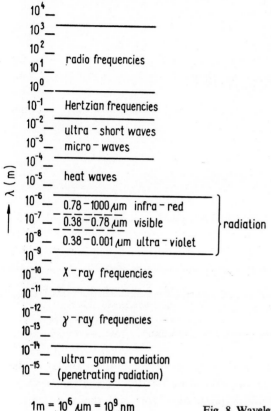

1m = 10^6 µm = 10^9 nm

Fig. 8 Wavelengths of the electromagnetic spectrum.

3.3 WAVE CHARACTERISTICS OF RADIATION

It has been proved both theoretically and experimentally that light radiation is a temporally and spatially periodic field, in which the components of the electric and magnetic intensity are related. The two waves (magnetic and electric) oscillate at equal phases in planes that are normal to one another and at the same time normal to the direction of the propagation (see Figure 9a). The phase is a temporal displacement of a wavelength with respect to the initial signal

Restricting ourselves to the electric field of the light wave [1, 2, 4], we can write the equation for the intensity of the field, **E**, of this light wave characterized by frequency v:

$$\mathbf{E} = A \cos 2\pi v\left(t - \frac{\mathbf{x}}{c}\right) = A \cos 2\pi(vt - k\mathbf{x}), \tag{40}$$

where A is the amplitude,
    t denotes the time,
    x the direction of the propagation,
    λ the wavelength,
    k the wave number,
    v the frequency.

The curve of the function is presented in Figure 9b.

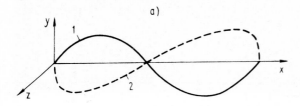

a)

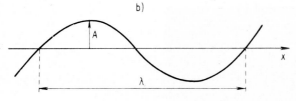

b)

Fig. 9 Wave characteristic of electromagnetic radiation: *1* — electric wave, *2* — magnetic wave.

In the natural, unpolarized light the electric waves oscillate at all angles to the direction of the propagation of the ray. If, however, the oscillations exhibit particular arrangements, we speak about polarized light; and the polarization can be either linear, elliptical, or circular. (This subject will be dealt with in Section 4.5.)

3.4   QUANTUM CHARACTERISTICS OF RADIATION

Since electromagnetic theory failed to describe the interaction of radiation with matter, Planck advanced his hypothesis assuming that radiant energy was emitted by bodies in the form of light quanta, which he called *photons*. The energy E of a light quantum is

$$E = hv, \tag{41}$$

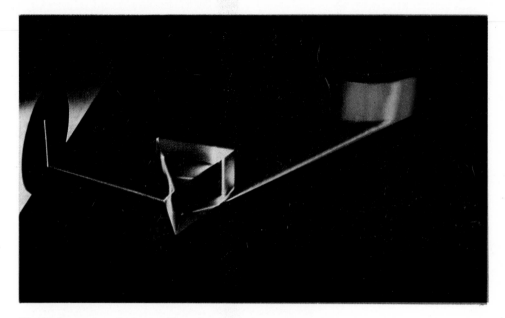

Fig. 12 Separation of a beam of light into a spectrum by glass prism [7].

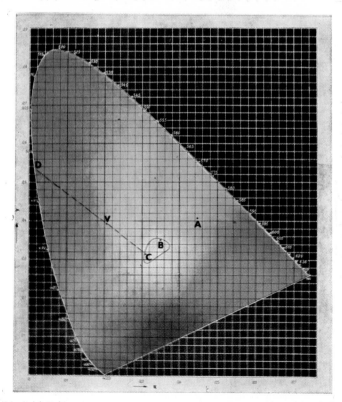

Fig. 15 Chromaticity diagram (colour triangle).

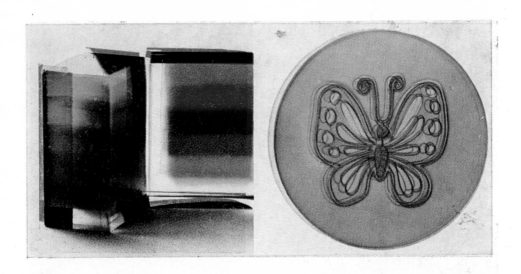

Fig. 193 Polychromatic glass.

and energy is always an integer multiple **n** of the light quantum [1, 3]

$$E = \mathbf{n}h\nu. \tag{42}$$

Later, Einstein proved that for the description of the photoelectric effect the assumption of the quantum nature of radiation is also inevitable. And he based his interpretation of photochemical effects upon the same principles. According to his definition, one light quantum acts upon one molecule in photochemical reactions. One molecule can thus absorb one light quantum, the result will, however, be an increase in the energy of this molecule by $h\nu$[3]. The energy absorbed by one mole of material N then equals

$$E = Nh\nu. \tag{43}$$

Expressing the wavelength in nm, we get relationship for the calculation of energy E of one mole of light quanta

$$E = \frac{1232.7}{\lambda} \quad [eV]. \tag{44}$$

The energy of one mole of light quanta expressed in joules makes

$$1 \text{ eV} = 1.6021 \cdot 10^{-19} \quad [J]. \tag{45}$$

Table 3 presents a list of the energies of one mole of light quanta expressed in joules and electron volts for the wavelengths from 250 to 800 nm.

Table 3

λ [nm]	E [eV]	E [J]
250	4.93	$7.898 \cdot 10^{-19}$
300	4.11	$6.585 \cdot 10^{-19}$
350	3.52	$5.639 \cdot 10^{-19}$
400	3.08	$4.934 \cdot 10^{-19}$
450	2.74	$4.390 \cdot 10^{-19}$
500	2.46	$3.941 \cdot 10^{-19}$
550	2.24	$3.589 \cdot 10^{-19}$
600	2.05	$3.284 \cdot 10^{-19}$
650	1.89	$3.028 \cdot 10^{-19}$
700	1.76	$2.820 \cdot 10^{-19}$
750	1.64	$2.627 \cdot 10^{-19}$
800	1.54	$2.467 \cdot 10^{-19}$

Radiation has a dual nature. In the experiments with interference, diffraction, and polarization it is the wave characteristic that is directly exhibited, whereas in photoelectric and photochemical effects, and in the fluctuations of the light field, it is the quantum property that becomes most apparent.

Light and heat sources invariably send out polychromatic (complex) radiation, even though this radiation may not be of equal intensity. These sources are therefore viewed as aggregates of micro-oscillators of various frequencies. The frequency responses of technical sources depend, for example, also upon the reflective properties of the surface. For this reason, a basic emitter such as an ideal black body or full radiator had to be introduced.

A *black body* is a perfect absorber of radiation, and the radiation that it emits is thus solely its own radiation.

According to Kirchhoff's definition, the intensity of the radiation emitted by black body (Planckian radiator) is dependent upon the absolute temperature of that body, thus

$$M_{e,\lambda} = f(T, \lambda), \tag{46}$$

where $M_{e,\lambda}$ denotes monochromatic emission.

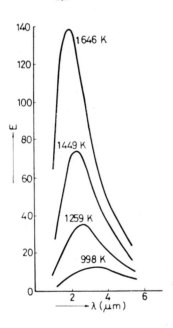

Fig. 10 Graphical representation of Planck's radiation law.

Stefan established that the maximum radiant intensity of a black body is dependent upon the wavelength. With temperature increasing, the maximum moves towards shorter wavelengths, viz. towards higher frequencies. The total radiant intensity of the source then increases with the fourth power of the temperature. Thus,

$$M_{e,\lambda} = \sigma T^4, \tag{47}$$

where T denotes absolute temperature,
and σ is the Stefan-Boltzmann constant $= 5.67 . 10^{-8} \, [\mathrm{W . m^{-2} . K^{-4}}]$.

34

Wien's law formulates an inverse proportionality between the position of the maximum of radiant intensity, λ_{max}, and absolute temperature, T

$$T\lambda_{max} = A, \tag{48}$$

where A is given by $2.898 \cdot 10^{-3} \, [m \cdot K]$.

Stefan's and Wien's laws could, however, be interpreted solely in terms of classical ideas. Planck's contribution to the solution of the problem gives the response curve of the black body as:

$$M_{e,\lambda} = \frac{c_1}{\lambda^5 (e^{c_2/\lambda T} - 1)}, \tag{49}$$

where λ denotes the wavelength,

$$c_1 = 2\pi hc^2 = 3.741 \cdot 10^{-8} \quad [W \cdot m^2],$$

$$c_2 = \frac{hc}{k} = 1.439 \cdot 10^{-2} \quad [m \cdot K],$$

h is the Planck's constant, $6.626 \cdot 10^{-34} \quad [J \cdot s]$,

k is the Boltzmann's constant, $1.380 \cdot 10^{-23} \quad [J \cdot K^{-1}]$,

and T is the absolute temperature $[K]$.

Planck's radiation law is in accordance with the results of experimental measurements given in Figure 10.

3.5   COMPLEX AND MONOCHROMATIC LIGHT

White daylight is an ideal source of light: in its spectrum are represented all the wavelengths (energies) of the visible region of the spectrum, their distribution being similar to that found in solar light. Figure 11 depicts the sun spectrum, including the ultra-violet and infra-red components.

If a ray of white light passes through a glass prism, it is refracted at different angles dependent upon the wavelength of the radiation. The result is a separation of the ray and a distribution of the waves into the order of their lengths, and hence, in accordance with Planck's law, in the order of the energies of the photons. Figure 12 represents the separation of white light by a glass prism into a spectrum [7].

We generally distinguish between two types of spectra: emission and absorption spectra.

Emission spectra are the result of radiation originating in atoms or molecules that have been brought into an excited state by added energy. While these atoms or molecules strive to regain their original ground state, the exciting energy consumed is emitted in the form of radiation.

Absorption spectra, on the other hand, are the result of absorption of certain light frequencies.

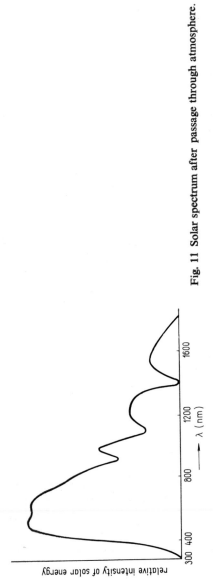

relative intensity of solar energy

λ (nm)

Fig. 11 Solar spectrum after passage through atmosphere.

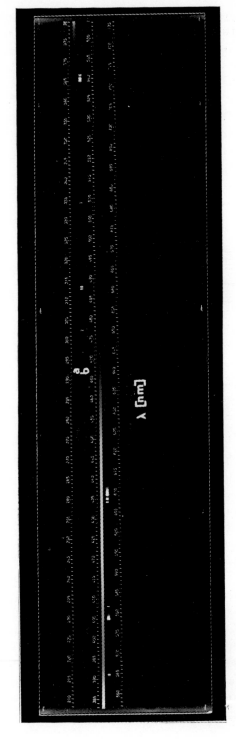

λ [nm]

Fig. 13 Continuous (a) and line (b) spectra.

According to the width of the absorption, or the emission lines, we distinguish between two types of spectra: the *line spectra*, arising from transitions between a limited number of levels (see Fig. 13b), and *continuous spectra*, arising from energy transitions among a large number of levels. For a wide range of the wavelengths (energies) of the spectrum we thus have more than one transition (see Fig. 13a).

Complex light is a blend of various wavelengths. The radiation defined by the wavelengths which are shorter or longer than those of the visible region of the spectrum cannot be perceived by the human eye, and is thus not referred to as light.

3.6   SPECTRAL SENSITIVITY OF THE HUMAN EYE

The sensitivity of the human eye to the different wavelengths of the visible region of the spectrum varies. Curve *1* in Figure 14 gives the relative sensitivity of the human eye, V_λ, the so-called photopic vision, in its dependence upon the wavelength, λ. We note that the sensitivity of the human eye declines very steeply toward the boundaries of the visible region of the spectrum, reaching its maximum at a wavelength of 555 nm. In Table 4 we present a list of the internationally standardized values of the relative luminous efficiency of the human eye, together with the values of luminous equivalent [lm . W^{-1}].

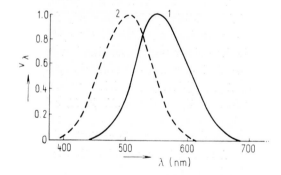

Fig. 14 Dependence of relative luminous efficiency of human eye, V_λ, upon wavelength: *1* — photopic vision, *2* — scotopic vision.

In Figure 14, curve *2* represents the luminous efficiency of the human eye in dim light, the so-called scotopic vision. We note that the sensitivity maximum has shifted in the direction towards the shorter wavelengths with the maximum at 507 nm. This phenomenon is the Purkyně effect, which makes the eye at daybreak take in blue hues first and only then the red ones. The so-called Bezold-Brücke effect is a term denoting variations of the colour sensation dependent upon the adaptation level in the region of daylight vision.

Table 4

λ [nm]	Relative luminous efficiency [V_λ]	Luminous equivalent [lm/W]	λ [nm]	Relative luminous efficiency [V_λ]	Luminous equivalent [l/m/W]
380	0.000 04	0.027	580	0.870	590.0
400	0.000 4	0.272	600	0.631	429.0
420	0.004	2.72	620	0.301	211.0
440	0.023	15.63	640	0.175	119.0
460	0.060	40.8	660	0.061	41.5
480	0.139	94.6	680	0.017	11.6
500	0.323	219.5	700	0.004 1	2.72
520	0.710	483.0	720	0.001 05	0.714
540	0.994	676.0	740	0.000 25	0.170
555	1.000	680.0	750	0.000 12	0.082
560	0.996	677.0			

3.7 MEASURING COLOURS

The Commission Internationale d'Eclairage (CIE) has adopted a special method of determining *chromaticity* or colour quality of any light stimulus or objects illuminated by a given light stimulus. Using this method, it is possible to induce the sensation of any colour by mixing additively three properly selected reference (basic) stimuli. The colorimetric magnitudes of the latter constitute a measuring scale with the help of which the colour under examination can be characterized in a colorimetric system. The CIE standard colorimetric system is defined by the spectral *tristimulus values* $\bar{x}(\lambda)$, $\bar{y}(\lambda)$, $\bar{z}(\lambda)$, which represent the relative colorimetric magnitudes of the reference light stimuli X, Y, Z, selected from an equi-energy spectrum [8].

The reference light stimuli agreed upon are monochromatic radiations of the $\lambda = 700$ nm wavelength for red light, the $\lambda = 546.1$ nm wavelength for green light, and the $\lambda = 435.8$ nm wavelength for blue light.

The tristimulus values $\bar{x}(\lambda)$, $\bar{y}(\lambda)$, $\bar{z}(\lambda)$ are applied, provided the angle of observation does not exceed 4°. If observations are made at angles larger than 4°, use must be made of *supplementary tristimulus values*.

Chromaticity can thus be simulated with the help of additively mixed magnitudes of three reference stimuli X, Y, Z; these magnitudes of reference stimuli are termed *trichromatic components of a light stimulus*. They can be calculated by integrating the product of the relative spectral composition of the reference chromatic light of a colour stimulus, $F(\lambda)$, and the tristimulus values $\bar{x}(\lambda)$, $\bar{y}(\lambda)$, $\bar{z}(\lambda)$. They are denoted as X, Y, Z, and the respective relationships are

$$X = \int F(\lambda)\, \bar{x}(\lambda)\, \mathrm{d}\,(\lambda),$$
$$Y = \int F(\lambda)\, \bar{y}(\lambda)\, \mathrm{d}(\lambda), \tag{50}$$
$$Z = \int F(\lambda)\, \bar{z}(\lambda)\, \mathrm{d}\,(\lambda).$$

Defining a colour sensation in full would thus involve making use of a three-dimensional representation in a colour space, X, Y, Z. Since spatial representation does not seem to be practical, a plane section of the colour space, viz. the *colour triangle* (chromaticity diagram) xy, is invariably resorted to.

The chromaticity coordinates can be calculated from the individual trichromatic components, viz.

$$x = \frac{X}{X + Y + Z}, \tag{51}$$
$$y = \frac{Y}{X + Y + Z},$$
$$z = \frac{Z}{X + Y + Z}.$$

Since $x + y + z = 1$, coordinate z can be omitted. Figure 15 shows the chromaticity diagram (colour triangle).

Whereas the chromaticity of primary sources is given by the spectral composition of their radiation, the chromaticity of secondary light sources does not depend solely upon the spectral transmittance or reflectivity of those sources, but also upon the spectral composition of the radiation of the source illuminating the secondary source. It is therefore essential that we specify the conditions for the measurement as far as the source applied is concerned, and the geometric conditions for the illumination and observation. Six basic light sources, the so-called *CIE conventional achromatic light stimuli*, are stipulated, the physical materialization of which has been determined.

For the four CIE standard achromatic light stimuli Table 5 quotes the chromaticity coordinates and the corresponding equivalent temperatures of chromaticity (the equivalent chromaticity temperature being the temperature of the Planckian radiator the radiation of which posseses chromaticity equal to the chromaticity of the radiation under consideration).

In the colour triangle each colour is thus defined by a point of intersection of chromaticity coordinates x, y. For instance, the line connecting the CIE standard achromatic light C (see Figure 15) with the point corresponding to colour V intersects the spectrum locus at point D, giving the so-called *dominant wavelength* (i.e., the wavelength of light that, mixed with achromatic light, will yield the light whose chromaticity is equal to the chromaticity of the light considered). With non-spectral, purple hues in the vicinity of the line of pure purples, the connecting line extends from the point for the colour of achromatic light (point C) to the side away from the point for the colour under consideration, and again the point of intersection of the line with the spectrum locus is sought. The so-termed *complementary wavelength* is quoted.

Table 5

Achromatic light stimulus (CIE standard)	Equivalent colour temperature		x	y	Source
	K	μrd			
A	2 854	350.3	0.4476	0.4075	Artificial illumination by incandescent lamp
B	4 800	208.3	0.3485	0.3518	Mean daylight with the prevailing component of direct solar radiation
C	6 500	153.8	0.3101	0.3163	Mean daylight free from direct solar radiation
E	The colour corresponds to equi-energy spectrum (physically realizable source)				

The *excitation purity*, p_e, is computed from the relationship

$$p_e = \frac{y - y_w}{y_s - y_w}; \qquad p_e = \frac{x - x_w}{x_s - x_w}, \tag{52}$$

where x, y are the chromaticity coordinates of the light considered,

x_s, y_s the chromaticity coordinates of the dominant wavelength of the light considered,

and x_w, y_w the chromaticity coordinates of the selected achromatic light.
The *colorimetric purity*, p_c, is computed from the relatoiship

$$p_c = p_e \frac{y_s}{y}. \tag{53}$$

In the colour space X, Y, Z, or in its plane section (colour triangle xy), equal linear distances in the different loci of the space, or triangle, do not correspond to equal subjective sensory differences of colour, and, vice versa, equal subjective sensory differences of colour do not correspond to equal distances in the different loci of the space, or triangle. Hence, a uniform colour space, or *uniform chromaticity scale diagram* (UCS), is extensively used (Fig. 16).

The orientation in a uniform chromaticity scale diagram is similar to the orientation in a colour triangle xy. The coordinates u, v of that diagram are calculated

40

either from the tristimulus colour values X, Y, Z, viz.

$$u = \frac{4X}{X + 15Y + 3Z}; \qquad v = \frac{6Y}{X + 15Y + 3Z}, \tag{54}$$

or from chromaticity coordinates x, y, viz.

$$u = \frac{4x}{-2x + 12y + 3}; \qquad v = \frac{6y}{-2x + 12y + 3}. \tag{55}$$

The coordinates x, y and u, v for the different spectral wavelengths in a colour triangle are listed in Table 6 and 7.

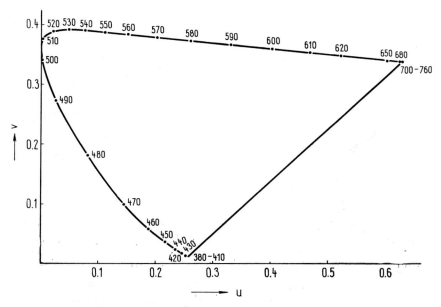

Fig. 16 Uniform chromaticity scale diagram.

Another method involves the application of the so-called *discriminating ellipse* (McAdams ellipse). Table 8 lists the values for the construction of the discriminating ellipses in the different regions of the colour triangle, viz. the transverse and the conjugate half axes, and the angle of inclination of the transverse axis to coordinate x.

The numerical values tabulated in Table 8 give an approximate fifth multiple of the change in the colour perceived. We are thus in a position to delineate in the form of an ellipse the colour region surrounding point x, y in a scale of the least colour deviations sensed. Figure 17 gives an example of a discriminating ellipse circumscribing colour $P(x = 0.35, y = 0.35)$. Since this is a graphical method, it yields ready visual information which the observer can immediately use in drawing conclusions concerning changes in colour.

Table 6

λ [nm]	x	y	z	λ [nm]	x	y	z
380	0.1741	0.0050	0.8209	595	0.6029	0.3965	0.0006
385	0.1740	0.0050	0.8210	600	0.6270	0.3725	0.0005
390	0.1738	0.0049	0.8213	605	0.6482	0.3514	0.0004
395	0.1736	0.0049	0.8215	610	0.6658	0.3340	0.0002
400	0.1733	0.0048	0.8219	615	0.6801	0.3198	0.0002
405	0.1730	0.0048	0.8222	620	0.6915	0.3083	0.0002
410	0.1726	0.0048	0.8226	625	0.7006	0.2993	0.0001
415	0.1721	0.0048	0.8231	630	0.7079	0.2920	0.0001
420	0.1714	0.0051	0.8235	635	0.7140	0.2859	0.0000
425	0.1703	0.0058	0.8239	640	0.7190	0.2809	0.0000
430	0.1689	0.0069	0.8242	645	0.7230	0.2770	0.0000
435	0.1669	0.0086	0.8246	650	0.7260	0.2740	0.0000
440	0.1644	0.0109	0.8247	655	0.7283	0.2717	0.0000
445	0.1611	0.0138	0.8251	660	0.7300	0.2700	0.0000
450	0.1566	0.0177	0.8257	665	0.7311	0.2689	0.0000
455	0.1510	0.0227	0.8263	670	0.7320	0.2680	0.0000
460	0.1440	0.0297	0.8263	675	0.7327	0.2673	0.0000
465	0.1355	0.0399	0.8246	680	0.7334	0.2666	0.0000
470	0.1241	0.0578	0.8181	685	0.7340	0.2660	0.0000
475	0.1096	0.0868	0.8036	690	0.7344	0.2656	0.0000
480	0.0913	0.1327	0.7760	695	0.7346	0.2654	0.0000
485	0.0687	0.2007	0.7306	700	0.7347	0.2653	0.0000
490	0.0454	0.2950	0.6596	705	0.7347	0.2653	0.0000
495	0.0235	0.4127	0.5638	710	0.7347	0.2653	0.0000
500	0.0082	0.5384	0.4534	715	0,6347	0.2653	0.0000
505	0.0039	0.6548	0.4313	720	0.7347	0.2653	0.0000
510	0.0139	0.7502	0.2359	725	0.7347	0.2653	0.0000
515	0.0389	0.8120	0.1491	730	0.7347	0.2653	0.0000
520	0.0843	0.8338	0.0919	740	0.7347	0.2653	0.0000
525	0.1142	0.8262	0.0596	745	0.7347	0.2653	0.0000
530	0.1547	0.8059	0.0394	750	0.7347	0.2653	0.0000
535	0.1929	0.7816	0.0255	760	0.7347	0.2653	0.0000
540	0.2296	0.7543	0.0161	770	0.7347	0.2653	0.0000
545	0.2658	0.7243	0.0099	780	0.7347	0.2753	0.0000
550	0.3016	0.6923	0.0061				
555	0.3373	0.6589	0.0038				
560	0.3731	0.6245	0.0024				
565	0.4087	0.5896	0.0017				
570	0.4441	0.5547	0.0012				
575	0.4788	0.5202	0.0010				
580	0.5125	0.4866	0.0009				
585	0.5448	0.4544	0.0008				
590	0.5752	0.4242	0.0006				

Table 7

λ [nm]	u	v	λ [nm]	u	v
380	0.2568	0.0111	595	0.3681	0.3631
385	0.2566	0.0111	600	0.4035	0.3596
390	0.2564	0.0108	605	0.4379	0.3561
395	0.2561	0.0108	610	0.4692	0.3530
400	0.2557	0.0106	615	0.4968	0.3503
405	0.2552	0.0106	620	0.5203	0.3479
410	0.2545	0.0106	625	0.5399	0.3460
415	0.2537	0.0106	630	0.5565	0.3443
420	0.2522	0.0113	635	0.5709	0.3429
425	0.2496	0.0128	640	0.5822	0.3412
430	0.2461	0.0151	645	0.5929	0.3407
435	0.2411	0.0186	650	0.6005	0.3400
440	0.2347	0.0233	655	0.6064	0.3394
445	0.2266	0.0291	660	0.6109	0.3389
450	0.2161	0.0366	665	0.6138	0.3386
455	0.2033	0.0459	670	0.6162	0.3384
460	0.1877	0.0581	675	0.6180	0.3382
465	0.1690	0.0746	680	0.6199	0.3380
470	0.1241	0.1007	685	0.6215	0.3378
475	0.1147	0.1362	690	0.6226	0.3377
480	0.0828	0.1805	695	0.6231	0.3377
485	0.0521	0.2285	700	0.6234	0.3377
490	0.0282	0.2745	705	0.6234	0.3377
495	0.0119	0.3123	710	0.6234	0.3377
500	0.0035	0.3420	720	0.6234	0.3377
505	0.0014	0.3621	730	0.6234	0.3377
510	0.0046	0.3759	740	0.6234	0.3377
515	0.0123	0.3846	750	0.6234	0.3377
520	0.0231	0.3891	760	0.6234	0.3377
525	0.0360	0.3908	770	0.6234	0.3377
530	0.0501	0.3912			
535	0.0643	0.2910			
540	0.0792	0.3904			
545	0.0953	0.3894			
550	0.1127	0.3880			
555	0.1319	0.3864			
560	0.1531	0.3844			
565	0.1766	0.3821			
570	0.2026	0.3796			
575	0.2312	0.2767			
580	0.2623	0.2736			
585	0.2960	0.3703			
590	0.3315	0.3667			

Tab. 8

Transverse half axis   Conjugate half axis

Angle of inclination

Each cell entry is given as: transverse value, small angle value / conjugate value.

λ \ x	0	0.05	0.10	0.15	0.20	0.25	0.30	0.35	0.40	0.45	0.50	0.55	0.60	0.65	0.70	0.75
0.85		89 24 / 95	88 25 / 92	86 26 / 88	77 23 / 85	69 21 / 82	62 19 / 77	56 17 / 73	51 16 / 69	46 14 / 64	41 12 / 59	37 10 / 54	33 8 / 49	30 7 / 44	27 5 / 39	28 5 / 37
0.80	83 23 / 99	81 23 / 95	80 23 / 92	78 23 / 88	70 22 / 85	63 20 / 81	56 18 / 77	50 16 / 72	45 14 / 67	40 12 / 62	36 11 / 57	32 9 / 52	29 7 / 47	26 6 / 42	22 4 / 37	23 4 / 35
0.75	75 22 / 99	73 22 / 96	72 22 / 92	71 22 / 88	63 20 / 85	56 18 / 80	50 17 / 76	45 15 / 71	40 13 / 66	36 11 / 60	32 9 / 55	28 8 / 49	25 6 / 44	22 5 / 39		
0.70	68 21 / 100	66 21 / 96	65 21 / 92	64 21 / 88	56 19 / 84	50 17 / 80	44 15 / 75	39 14 / 69	35 12 / 63	31 11 / 58	27 8 / 52	24 7 / 46	21 5 / 41			
0.65	61 20 / 100	59 20 / 97	58 19 / 93	57 19 / 88	50 18 / 84	44 16 / 79	39 14 / 73	35 12 / 67	30 11 / 61	26 9 / 55	23 7 / 49	20 6 / 43	17 4 / 38			
0.60	54 18 / 101	53 18 / 97	51 18 / 93	51 18 / 88	44 16 / 84	39 15 / 78	34 12 / 72	30 11 / 65	26 10 / 58	22 8 / 52	20 6 / 46	17 4 / 40				
0.55	48 17 / 102	47 17 / 98	45 17 / 94	45 17 / 89	39 15 / 83	34 14 / 77	29 12 / 70	25 10 / 62	22 8 / 55	19 6 / 49	16 5 / 42					
0.50	43 16 / 103	41 16 / 99	40 16 / 94	39 16 / 89	34 14 / 83	29 12 / 76	25 11 / 67	21 9 / 59	18 7 / 52	16 5 / 45	13 4 / 39					
0.45	37 15 / 104	36 15 / 100	35 15 / 95	34 14 / 89	29 13 / 82	24 11 / 73	21 10 / 64	18 8 / 55	15 6 / 48	12 4 / 40						
0.40	32 14 / 106	31 14 / 102	30 13 / 96	29 13 / 89	24 12 / 80	20 10 / 70	17 8 / 60	14 6 / 51	12 5 / 43							
0.35	28 13 / 109	26 12 / 103	25 12 / 97	24 12 / 89	20 10 / 78	17 9 / 66	14 7 / 55	12 5 / 46	9 4 / 37							
0.30	23 12 / 111	22 11 / 105	21 11 / 98	20 11 / 89	17 9 / 75	14 8 / 61	11 6 / 50	9 5 / 40								
0.25		18 10 / 108	17 10 / 100	16 10 / 87	13 8 / 70	11 6 / 55	9 5 / 49									
0.20		14 9 / 111	13 9 / 101	13 8 / 85	10 7 / 64	8 5 / 48	6 4 / 40									
0.15		11 8 / 115	10 8 / 103	10 7 / 84	8 6 / 57	6 4 / 40										
0.10			8 7 / 104	8 6 / 80	6 4 / 48											
0.05			6 6 / 105	6 5 / 75	5 3 / 38											
0				5 4 / 68												

→ x

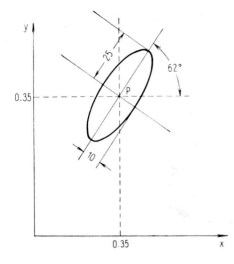

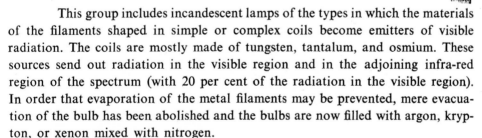

Fig. 17 Discriminating ellipse in a colour triangle circumscribing colour P.

The coordinates of colour triangles, x, y (Table 6), and the transverse and the conjugate half axes (Table 8) are in relation 1 : 1000.

3.8 SOURCES OF LUMINOUS RADIATION

The basic luminous sources of radiation can be roughly classified, according to [4, 18], into five main groups: a) thermal sources, b) arc lamps, c) discharge tubes, d) luminescent sources, and e) lasers.

Sources of β-radiation and γ-radiation, which cannot be viewed as clean sources of luminous radiation, are dealt with in Section 6.4 in the context of the discussion of the interaction of these types of radiation with glasses.

The sources of radiant energy generally produce either a continuous spectrum (see Fig. 13a) or more or less narrow emission lines (Fig. 13b).

3.8.1 Thermal sources of radiation

This group includes incandescent lamps of the types in which the materials of the filaments shaped in simple or complex coils become emitters of visible radiation. The coils are mostly made of tungsten, tantalum, and osmium. These sources send out radiation in the visible region and in the adjoining infra-red region of the spectrum (with 20 per cent of the radiation in the visible region). In order that evaporation of the metal filaments may be prevented, mere evacuation of the bulb has been abolished and the bulbs are now filled with argon, krypton, or xenon mixed with nitrogen.

For the infra-red region of the radiation use is also made of sources with silit or globar resistance-heated rods, or the Nernst glower (zirconium dioxide mixed with ytrium oxide and the rare earths), etc.

3.8.2   Arc lamps

From the point of view of the material that the electrodes are made of, and of the density of the current, there are three basic types of arc lamp sources of radiation:

(a) Arc lamps with pure carbon electrodes. Since the burning process in these lamps virtually involves carbon only, the glowing arc has the character of a grey body. With the enclosed arc, the short-wave and the infra-red part of the radiation manifest higher intensity.

(b) Arc lamps the anodes of which are made of carbon mixed with calcium, cerium, titanium, strontium, and other fluorides. These lamps send out selectively visible radiation.

(c) Arc lamps the carbon anodes of which contain additions of Ca, Ce, Ti, Sr, and of other fluorides, the cathodes being made of pure carbon. The sources of this type send out radiation closely resembling daylight. Tungsten, or other materials, can also be used for the electrodes.

3.8.3   Discharge tubes

The kind of gas or vapour, the pressure, the temperature, the type and structure of the electrodes, and other parameters, determine the conditions for electric discharge in these sources of light and thus the qualitative properties of the radiation. The approximate classification here is into:

(a) Low-pressure discharge tubes filled with rare gases (krypton, xenon, helium, and others, pure or mixed; also hydrogen or deuterium). The electrodes are either cold or heated.

(b) High-pressure discharge lamps filled with rare gases (xenon).

(c) Discharge lamps filled with metallic vapour: low-pressure tubes with cold electrodes (mercury filling), low-pressure tubes with heated electrodes (mercury, sodium filling), medium-pressure tubes with heated electrodes (containing inert gas with a defined quantity of mercury), the pressure ranging from 0.5 to $10 . 10^5$ Pa, and high-pressure tubes (filled with inert gas with a defined quantity of mercury), the pressure ranging between 10 and $304 . 10^5$ Pa.

The most current sources of this type are hydrogen discharge lamps sending out intensive continuous radiation of the $200-380$ nm waveband and producing spectrum lines in the visible region.

The deuterium discharge lamps emit continuous radiation in the ultra-violet region; as compared with the hydrogen discharge tube the radiation is much more intensive.

A most frequently used source of intensive radiation is the mercury discharge lamp serving as a source of non-continuous radiation. These discharge lamps produce lines of higher or lower intensity ranging from the ultra-violet to the

46

infra-red region of the spectrum according to whether they are of the high-pressure, medium-pressure, or the low-pressure type.

Adding cadmium into the mercury discharge lamps raises the proportion of red light, but reduces the intensity of the emission.

The sodium discharge lamp produces spectrum lines in the visible region of the spectrum. It is filled with neon in order that the discharge of sodium may be facilitated; being heated, the sodium will evaporate and the igniting neon discharge will be followed by a sodium discharge.

Use is also made of potassium, helium, rubidium, thallium, zinc, and a number of other types of discharge lamps.

3.8.4   Luminescent sources of radiation

These sources are electroluminescent or radioluminescent, but their parameters can in neither case compare with the parameters of the classical sources dealt with in Sections 3.8.1 and 3.8.3.

An electroluminescent source of radiation consists of a zinc sulphide electro-luminophore (other materials can be used, too, e.g., carbon silicide, gallium monoarsenide, mixed sulphides, selenides, and tellurides of zinc, cadmium, and mercury) placed on a glass plate coated with a conductive layer of tin dioxide and sealed in with insulating varnish. The other electrode is a metal plate. The emission of zinc sulphide, according to how much it has been doped, is within the range of $400 - 700$ nm.

The radiation of a radioluminescent source is emitted by the isotope of krypton Kr^{85} or of tritium H^3. These isotopes, placed in a glass flask the surface of which has a luminophore coating, emit β- and γ-radiation; their half life is about ten years. Due to the extremely low intensity of their radiation, these sources can, however, be applied only in signalling devices and traffic signs etc.

3.8.5   Lasers

The quantum generator of light, the laser, is a source of highly mono-chromatic coherent radiation of the optical, mostly visible to infra-red, region of the spectrum. In contradistinction to the more current sources of light, the emission in this case is stimulated (see Chapter 2). The condition for the preponderance of stimulated emission over absorption is a state of inversion.

Inversion can be achieved in the non-stationary state by the exciting (pumping) energy, e.g., luminous energy, being introduced into the system (resonator) with the help of a xenon discharge lamp. Thus, if an electromagnetic wave passes through an inversion medium (the frequency of the passing radiation being considered equal to Bohr's frequency v for quantum transition), its intensity will not attenuate; on the contrary, it will amplify, and the preponderance of stimulated emission over absorption will be established.

The individual quantum systems will emit their photons in such a way that the direction of the propagation of these photons, and their phases, will coincide with the electromagnetic radiation in transit. As soon as the preponderance of emission over absorption has thus been established, and the secondary losses in the system overcome, the system will act as a generator of stimulated emission.

Current use is made of gas, crystal, semiconductor, and glass resonators. The latter are discussed in Section 6.6.

3.9   BASIC QUANTITIES IN THE FIELD OF LIGHT
       AND RADIATION

The inclusion of the light and radiation units into the International system of measures (SI) is not a result of scientifically based deliberation and decision but is much more the result of the recommendations and resolutions of general conferences on weights and measures and of the legalizing acts of the individual countries [6]. According to [9, 10. 12], these units are defined in two categories.

3.9.1   Photometric terms

Luminous flux Φ [1m]. A *lumen* is the magnitude of luminous flux emitted into a solid angle of 1 sr by a point source the luminous intensity of which equals 1 cd in all directions.

Luminous quantity Q_v [1ms] is the term for luminous flux per unit of time.

Illumination (illuminance) E_v [lx] is the ratio of the part of luminous flux incident upon an area of the surface of a body to the dimensions of that area.

Light exposure H [lxs] is the term for illumination per unit of time.

Luminous intensity I_v [cd] is the ratio of luminous flux directionally emitted by a source into a infinitely small solid angle to the degree of the inclusion of that angle.

Luminance L_v [nt] is the ratio of luminous intensity of an area of the surface of a source in a given direction to the projection of that area on a plane normal to the given direction.

3.9.2   Energy terms

Radiant flux Φ_e [W] is the power transmitted by electromagnetic radiation.

Irradiance E_e [W . m^{-2}] is the ratio of radiant flux incident upon a surface to the dimensions of that surface.

Radiant exposure He [W . s . m^{-2}] is the term for irradiance per unit of time.

48

Radiant intensity I_e [W . sr^{-1}] is the ratio of radiant flux emitted by a point source in a cone of an infinitely small solid angle the axis of which lies in the direction of the radiation considered to the degree of inclusion of the solid angle of the cone.

Radiance (radiant intensity per unit area) L_e [W . m^{-2} . sr^{-1}] is the ratio of radiant intensity of an area of the surface of a source in a given direction to the projection of that area on a plane perpendicular to the given direction.

Further quantities and units are quoted in the works listed in the references [6, 7, 9, 10, 11, 12, 13, 17] at the end of this chapter.

References

[1] M. V. Volkenštejn, Struktura a fyzikální vlastnosti molekul (Structural and Physical Properties of Molecules). NČAV, Prague 1962.

[2] M. G. Veselov, Úvod do kvantové teorie atomů a molekul (Introduction to Quantum Theory of Atoms and Molecules). SNTL, Prague 1966.

[3] J. Kubal, Základy fotochemie (Foundations of Photochemistry). Academia, Prague 1969.

[4] V. Kmeť, Svetlo a osvetlovacie zariadenia (Light and Lighting Equipment). SVTL, Bratislava 1961.

[5] M. V. Volkenštejn, Molekularnaya optika (Molecular Optics). GITTL, Moscow 1951.

[6] M. Brezinščák, Veličiny a jednotky v technické praxi (Quantities and Units in Technology). SNTL, Prague 1970.

[7] Committee on Colorimetry of the Optical Society of America: The Science of Color. T. Y. Crowell Co., New York 1953.

[8] ČSN 01 1718 (Czechoslovak Standard) — Měření barev (Measuring Colours). 1966.

[9] ČSN 01 1701 (Czechoslovak Standard) — Základní veličiny v oboru světla a záření (Basic Quantities in the Field of Light and Radiation). 1954.

[10] ČSN 01 1711 (Czechoslovak Standard) — Světelné jednotky (Light Units). 1954.

[11] J. Binko, Fyzikální a technické veličiny (Physical and Technological Quantities). SNTL, Prague 1968.

[12] ČSN 01 1300 (Czechoslovak Standard) — Zákonné měrové jednotky (Czechoslovak Standard Units of Measure). 1974.

[13] Mezinárodní soustava jednotek SI (International System of SI Units). VÚNM, 1973.

[14] M. Malát, Absorpční anorganická fotometrie (Absorptive Inorganic Photometry). Academia, Prague 1973.

[15] A. N. Zajdel, V. K. Prokofev, and S. M. Rajskij, Tablici spectralnykh linyi (Tables of Spectral Lines). GITTL, Moscow 1952.

[16] ČSN 01 1710 (Czechoslovak Standard) — Poměrná světelná účinnost jednobarevného záření (Relative Luminous Efficiency of Monochromatic Radiation). 1954.

[17] ČSN 36 0000 (Czechoslovak Standard) — Světelně technické názvosloví (Terminology of Light Technology). 1967.

[18] S. Miškařík, Moderní světelné zdroje (Modern Light Sources). SNTL, Prague 1979.

[19] W. Heitler, The Quantum Theory of Radiation, Univ. Press, London—New York 1954.

[20] S. G. Lipson and H. Lipson, Optical Physics. Cambridge University Press, Cambridge 1966.

[21] J. M. Stone, Radiation and Optics, McGraw-Hill, New York 1963.

[22] M. Born and E. Wolf, Principles of Optics. Pergamon Press, New York 1959.

Since we are now going to deal with the interraction of radiation with glass, we consider it logical that we should give at least an introductory brief definition of the vitreous state. The definition should possess general validity and it should offer a description of the vitreous state, regardless of the possible structural changes e.g., of the growth of a crystalline phase, phase separation caused by heat treatment, irradiation, or other external effects.

Amorphous and vitreous state

Glasses are generally defined as materials in an amorphous state viz. as materials not manifesting properties typical of crystal lattices [24]. Figure 18 gives the X-ray diffraction patterns of crystalline quartz and amorphous quartz glass.

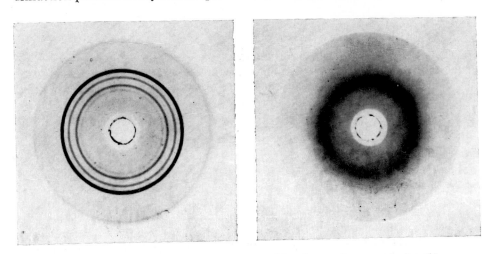

Fig. 18 X-ray diffraction patterns of crystalline quartz (a) and amorphous quartz glass (b).

Thus, within the limits of the accuracy of the X-ray diffraction method, glasses do not manifest the same periodicity or symmetry of structural regions to large distance (Fig. 18b) as crystalline materials (Fig. 18a).

Cooling the melt below the melting point, or, with multi-component systems, below the liquidus temperature, will not produce crystallization but only super-cooled liquid. With a given pressure, temperature, and composition, an equilibrium state (metastable equilibrium) will establish itself after some time corresponding to a certain configuration of the molecules in the liquid [10]. The establishment of the equilibrium state always lags behind changes in the temperature or pressure, because setting up the equilibrium configuration of the molecules requires a certain so-called relaxation time. In some systems the relaxation time increases rapidly with decreasing temperature, so that at a certain point the rate of cooling will be higher than the rate at which the equilibrium configuration is established. And the equilibrium configuration corresponds to that temperature: it seems to have "frozen" and will not change even if cooling continues. The super-cooled liquid has thus passed into the solid phase named glass. The temperature at which "freezing" is effected is the temperature of transformation, T_g, which depends upon the rate of cooling and does thus not constitute a characteristic material constant (see Fig. 19).

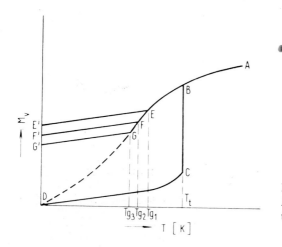

Fig. 19 Dependence of molar volume, M_v, upon temperature in the vicinity of the melting point for a single-component system.

Segment $\overline{AB}$ in Figure 19 is the liquid phase; the abrupt change in the volume, $\overline{BC}$, represents the conversion into a crystalline material (solidification temperature or melting temperature T_t), the volume of which subsequently changes with temperature along $\overline{CD}$.

$\overline{BD}$ represents the equilibrium dependence of the molar volume, M_v, of the super-cooled liquid upon temperature (the dashed part being a hypothetical region); segments $\overline{EE'}$, $\overline{FF'}$, and $\overline{GG'}$ visualize the temperature-dependence of the molar volume of glasses of different thermal histories. The rate of cooling has been the highest with glass $\overline{EE'}$. The properties of glasses are thus dependent upon the rate of cooling, as can be seen in Figure 19; and the range of temperatures defined by T_{g1} and T_{g3} is the so-called transformation range.

Glass is thus a solid, non-crystalline (amorphous) phase. The upper limit of the region of its existence is limited by the so-called transformation temperature,

T_g, or pressure, at which glass passes into a super-cooled liquid. Since the transformation temperature is not a material constant, in order that crystallization may be prevented, a certain rate of cooling must be applied. It is therefore logical that we should have dealt with the so-called kinetic theory of the formation of glass in this chapter. These problems have been discussed in more detail by Turnbull, Cohen, Hillig, Uhlmann and others [11, 12, 13, 14].

The rate of the formation of the nuclei (nucleation rate) can, according to Turnbull [11], be expressed as follows:

$$I = n''\nu\, e^{-\frac{N_A W^*}{kT}}\, e^{-\frac{\Delta G'}{kT}}, \tag{56}$$

where I is the number of stable nuclei formed in a unit volume of the system per unit of time,

  n'' the number of atoms in a unit of volume,
  ν the vibration frequency of the atoms at the boundary between nucleus and liquid,
  N_A the Avogadro number,
  T the temperature,
  k the Boltzmann constant,
and W^* the thermodynamic nucleation barrier given by the relationship

$$W^* = \frac{K\sigma'^3 V_m^2}{\Delta G^2}, \tag{57}$$

where K is a numerical value dependent upon the form of the nucleation (for spherical formation $K = 16\pi/3$),

  σ' the free inter-phase energy at the boundary between crystal and liquid,
  ΔG the change in free enthalpy during the passage of the liquid into crystal at temperature T,
and V_m the volume of one gram atom of crystal.

The exponent $e^{-\frac{N_A W^*}{kT}}$ gives the probability of occurrence of a nucleus of supercritical size (stable) at temperature T, and the exponent $e^{-\frac{\Delta G'}{kT}}$ the rate of the change of the structure of the material during nucleation. It includes the process of diffusion in the reorientation of the molecules ($\Delta G'$ giving the change in free enthalpy involved in overcoming the kinetic nucleation barrier).

If it happens that

$$\Delta G = \Delta T\, \Delta S = \Delta T\, \frac{\Delta H_t}{T_t}, \tag{58}$$

where ΔH_t is the change in enthalpy associated with melting,

  T_t the melting temperature,
and ΔS the entropy of crystallization,

52

then for the rate of nucleation at temperature T we get following relationship,

$$I = n''ve^{-\frac{N_A K \sigma'^3 V_m{}^2 T_t{}^2}{kT\Delta H_t{}^2 \Delta T^2}} e^{-\frac{\Delta G'}{kT}}. \tag{59}$$

With decreasing temperature, the nucleation rate does not however grow ad infinitum. The second exponential term gradually becomes dominant, and with further drop in the temperature the rate decreases (see Figure 20 after Rawson [8]).

The temperature at which the nucleation rate is maximal depends upon $\Delta G'$. If $\Delta G'_{T_t} = 0$, the nucleation rate reaches its maximum at temperature $T = \frac{T_t}{3}$, and the higher the value of $\Delta G'$, the more does this value approximate to the melting temperature T_t.

After a stable nucleus has been formed, the rate of its growth will depend upon the mobility, of the atoms diffusing towards its surface. Further, new bonds will be formed depending upon the structure of the crystal being formed. Figure 21 (after Turnbull [12]) presents the curve showing the variations of the free energy of the atoms at the boundary between the liquid phase and the crystal.

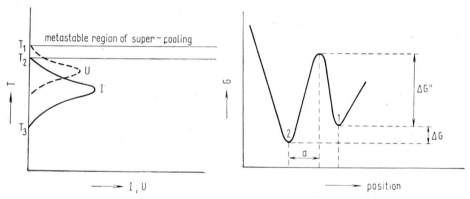

Fig. 20 Dependence of the rates of nucleation, I, and crystallization, U, upon super-cooling.

Fig. 21 Free energy diagram for a crystal.

It is assumed that atoms in position 1 (liquid phase) manifest higher energy than atoms in position 2 (crystal), the energies of the two positions differing by ΔG, which equals the free energy of crystallization. Surmounting the energy barrier (liquid-crystal boundary) requires activation energy $\Delta G''$, and, vice versa, overcoming the crystal-liquid boundary requires activation energy $\Delta G'' + \Delta G$.

The rate of growth of the crystal is thus proportional to the difference between the frequencies of the liquid-crystal and crystal-liquid transitions, v_{ls} and v_{sl}, viz.

$$v_{ls} = ve^{-\frac{\Delta G''}{kT}}, \tag{60}$$

$$v_{sl} = ve^{-\frac{\Delta G'' + \Delta G}{kT}}. \tag{61}$$

53

The rate of the growth of the crystal is then given by the relationship

$$U = a_0(v_{ls} - v_{sl}) = a_0 v \, e^{-\frac{\Delta G''}{kT}} \left(1 - e^{\frac{\Delta G}{kT}}\right), \tag{62}$$

where a_0 is the dimension of the molecule.

The growth of a crystal of a complex material is influenced by the rate of the diffusion towards its surface from relatively large distances. On the other hand, the rate of the nucleation is influenced solely by the activating energies associated with particles at short range.

For the simplest materials Turnbull and Cohen [11, 12] postulate that the activating energy controlling the rate of the growth of crystals should equal the activating energy governing viscous flow. Substituting into equation (62) we get

$$U = \frac{fkT}{N_A \, 3\pi a_0^2 \eta} \left[1 - e^{\frac{\Delta G}{kT}}\right], \tag{63}$$

where f is a factor between 0 and 1 (equal to the number of points of the surface of the crystal with which the molecules connect).

If cooling the melt is to produce glass, the rates of nucleation and crystallization must not exceed a certain value. If T' is the temperature and $\Delta T'$ the super-cooling at which $I = 1 \; [\text{cm}^{-3} . \text{s}^{-1}]$, then glass forms, according to Turnbull and Cohen [11, 12], if

(a) the nucleation rate does at no temperature T below the melting point T_t reach the value $1 \; [\text{cm}^{-3} . \text{s}^{-1}]$,

(b) at temperature T the rate of growth of the crystals is lower than 10^{-5} atoms per second.

Substituting $I = 1 \; [\text{cm}^{-3} . \text{s}^{-1}]$ into equation (59), we get

$$\frac{\Delta T'}{T_t} = \left[\frac{N_A K \sigma'^3 V_m^2}{\Delta H_t^2 (kT \ln(n''v) - \Delta G')}\right]^{1/2}. \tag{64}$$

With a number of materials the heat of crystallization, ΔH_t, is proportional to the melting temperature,

$$\Delta H_t = \beta k T_t, \tag{65}$$

where β approximates to 1 with the simpler liquids (molecular entropy of fusion in units of $\mathbf{k}$).

Substituting into equation (64) and applying the equation

$$\sigma' = \frac{\alpha \, \Delta H_t}{N_A^{1/2} V_m^{2/3}}, \tag{66}$$

(where α equals $1/3 - 1/2$),
we get

$$\frac{\Delta T'}{T_t} = \left[\frac{K \alpha^3 \beta k T_t}{kT' \ln(n''v) - \Delta G'}\right]^{1/2}. \tag{67}$$

If $\Delta G' = 0$ and the nuclei are spherical $(K = 16\pi/3)$, $n'' = 10^{23}$ [atoms/cm³], $v = 10^{12}$ [s⁻¹], we can write

$$\frac{\Delta T'}{T_t} = \left[\frac{16\pi}{3.80} \alpha^3 \beta \frac{T_t}{T'} \right]^{1/2}. \tag{68}$$

If $\alpha\beta^{1/3} \geqq 0.9$, a value can be found for T' satisfying equation (68). Pure materials crystallizing while cooled do not usually manifest $\alpha\beta^{1/3}$ greater than 0.5. If $\alpha\beta^{1/3} = 0.5$, and applying equation (67), we get

$$\frac{\Delta T'}{T_t} = \left[\frac{2\pi k T_t}{3.80 \cdot kT' - \Delta G} \right]^{1/2}. \tag{69}$$

For I not to exceed 1 [cm⁻³ . s⁻¹] for any temperature below T_t, the condition $\Delta G' \geqq 40\,kT_t$ must be satisfied.

In current glasswork practice the rates of cooling are generally so high that far higher rates of growth of crystals can be admitted. For instance, for the maximum rate of growth of crystals 0.1 [μm . s⁻¹] the maximum admissible values of $\Delta G'$ and $\Delta G''$ will equal $20\,kT_t$.

The structure of glasses is dealt with in detail in [1, 10, 15, 16] of the references.

Owing to the great variety of views describing the structure of glasses, no generally acceptable structural model has as yet been suggested. Due to the vast range of the groups of materials (elements, oxides, chalcogenides, "metal glasses", amorphous layers etc.) capable of glass formation after their melts have been cooled down to temperatures below the melting point or liquidus temperature (or if subjected to special technological treatment), the concept of a generalized structural model can hardly be expected to materialize; it can safely be assumed that the structures of the different groups of materials in the vitreous state are at least partially different from each other.

With the amount of information concerning the structure of glasses (which have so far been looked upon as isotropic materials in an amorphous state) increasing, it has become obvious that from the sub-microscopic point of view a number of these materials can unequivocally not be considered isotropic.

With the help of up-to-date measurement techniques, micro-heterogeneous centres (fluctuations of concentrations) have been discovered in glasses manifesting characteristics similar to immiscible liquids (i.e. no crystallization). Phase separation can therefore be observed in a number of cases (Fig. 22).

Distinction must be made between the so-called stable separation above the liquidus temperature and metastable separation below the liquidus temperature. [28, 30]. In view of the aim of the present book we shall briefly discuss metastable separation below the liquidus temperture.

In multi-component systems which have been supercooled below the liquidus temperature, the melt should be forming crystals. If it is in an amorphous state, this state is metastable, as already mentioned above. If cooling produces struc-

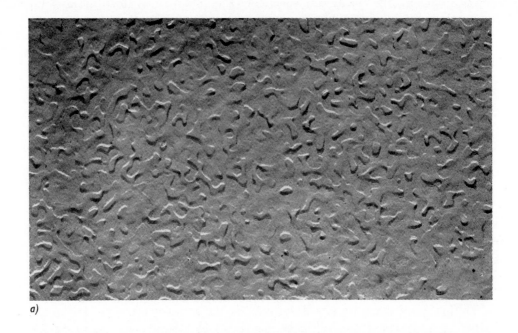

a)

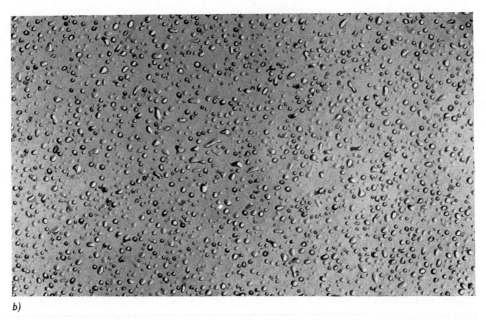

b)

Fig. 22 Phase separation in glass: spinodal decomposition (a), separation by nucleation and growth of separated phase (b).

56

tural clusters of a more stable phase (configurations more advantageous from the point of view of energy) of various dimensions but variable composition, separation can in general occur within certain temperature ranges. Thermal treatment within the region of such temperatures will make the separated phase grow.

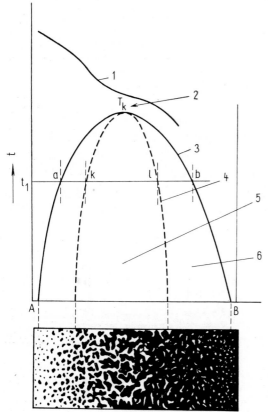

Fig. 23 Schematic diagram of phase separation below liquidus temperature in a binary system $A—B$: 1 — liquidus curve; 2 — critical temperature; 3 — binodal line; 4 — spinodal line; 5 — spinodal decomposition; 6 — nucleation and growth of separated phase.

In the simplest case, metastable separation below the liquidus temperature can proceed in two ways, according to Cahn, Charles, and Hilliard [17, 18, 19]:

(a) Let us first consider the phase diagram of a binary system $A - B$ (Fig. 23) defining a region of limited miscibility. For temperature t_1 we can in this case express, according to [10], the dependence of free energy, G, upon composition by the curve presented in Figure 24.

The stable system, viz. the system with the lowest free energy, will consist of two phases the composition of which is given by points a, b. The way in which the given system gets into this state depends upon the initial composition of that system. If the system x is subjected to abrupt cooling to temperature t_1, decomposition to phase a, b is initiated by way of a continuous change in the composition, because decomposition is linked with a decrease in free energy, ΔG, e.g., decomposition to phases 1, 1' (Fig. 24), as shown in Figure 25.

A continuous process of this type is termed *spinodal*; it manifests a structure consisting of two continuous interpenetrating phases (Figs. 22, 23). According to Cahn's theory spinodal decomposition occurs when

$$\frac{\partial^2 \Delta G}{\partial c^2} < 0. \tag{70}$$

The region of spinodal decomposition is divided by a *spinodal line* separating the unstable region of separation from the metastable region, viz.

$$\frac{\partial^2 \Delta G}{\partial c^2} = 0. \tag{71}$$

(b) Things are different if the initial composition is in the region where the curve of free energy is convex (see. Fig. 24). For the system of composition y decomposition to phase 2, 2' will be accompanied by an increase in free energy, $\Delta G'$, as visualized in Figure 26, and is thus not spontaneous.

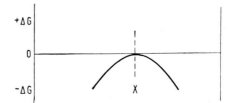

Fig. 25 Change in free energy ΔG associated with the decay of composition X.

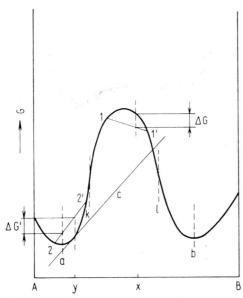

Fig. 24 Dependence of free energy, G, upon composition in binary system A—B for temperature t_1.

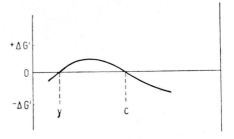

Fig. 26 Change in free energy ΔG associated with the decay of composition Y.

This region is termed *metastable*. If the structural clusters reach a certain critical radius depending upon the value of the inter-phase energy, they become *nuclei* of a new phase and keep *growing* to form a dispersive system. They are the so-called discrete particles dispersed in the second continuous phase (see Figures 22b and 23). For the region of metastability it is clear that

$$\frac{\partial^2 \Delta G}{\partial c^2} > 0. \tag{72}$$

The boundary between the two mechanisms of separation is marked by a point of inflexion k, l on the spinodal line. In the region endorsed by the spinodal line the separation of the phase is effected spinodally, in the region between the *spinodal* and the *binodal lines* nucleationally and by the occurrence of a new phase.

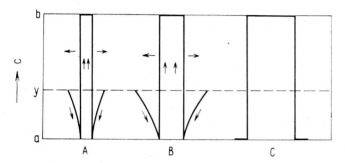

Fig. 27 Development of concentration profile during growth of a nucleated phase: A — at the moment of the appearance of a fluctuation in concentration; B — in the process of separation; C — in equilibrium.

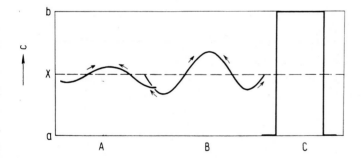

Fig. 28 Development of concentration profile during growth by spinodal decomposition. (Notation as in Figure 27).

The more the composition of the initial phase (super-cooled liquid) approximates to the spinodal line, the more diffuse will be the boundary between the nucleus and the parent phase, and the more continuous will be the transition of the composition of nucleation into the composition of the parent phase (for a schematic representation see Figure 23). The rise of the concentration profile at the boundary of the newly created phase is sketched in Figure 27 (after [10]). For spinodal decomposition the concentration profile is shown in Figure 28.

Another theory is Haller's kinetic theory [20, 21], according to which a high degree of continuity can be achieved through gradual coalescence of the individual drops of the separated phase.

After the initial state of separation, during the process of thermal treatment, the system strives to reduce the area of its interfacial surfaces by increasing the separated surfaces. Apart from coalescence, Brownian motion, or viscous flow, this process can be explained as the so-called Ostwaldian process of maturation, which is based upon the difference in the solubility of the smaller and the larger microregions of separation. The smallest regions, smaller than a certain value of critical radius r^*, have a tendency to dissolve, and to diffuse the dissolved ions to larger (more stable) regions, which then grow [22]. Figure 29 gives a schematic representation (after [22]) of the development of the concentration profile during the Ostwaldian process of maturation.

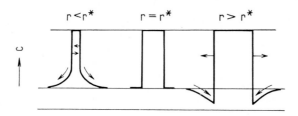

Fig. 29 Development of concentration profile during Ostwaldian maturation process.

If $r < r^*$ and if the concentration on the surface of the separated phase of radius r is higher than the concentration of the surrounding matrix, the phase will get dissolved in that matrix. If the concentration on the surface of the separated phase of radius r equals the concentration of the matrix, we have the case of an equilibrium of the two concentrations, and if $r > r^*$ and the concentration on the surface of the separated phase of radius r is lower than the concentration of the surrounding matrix, the separated phase is capable of growing.

In a number of cases it has been ascertained that further separation can even appear in the already separated phase.

Generally speaking, the problems discussed so far are rather complex, but the separation effect observed in glasses has greatly contributed to the solution of a number of questions concerning the structure of the vitreous systems [36].

Treating the subject below, we will in general consider glasses to be amorphous, isotropic materials, and the divergence from the assumed structural arrangement (or anisotropy) due to the chemical composition or to some external influence will be taken account of individually as the case may be.

Glasses passing from the liquid phase (melt) into the solid phase due to cooling, or vice versa, manifest thermal history, which influences the physico-chemical properties and, consequently, the optical properties of the glasses as well.

Interaction of radiation with glass

When radiation interacts with glass in the simplest way [2, 3, 35]

(a) it is reflected at the interface between the glass and the medium of incidence of the radiation if the refractive indices of the two media differ from one another;

(b) it is refracted at the interface between the two media manifesting differing indices of refraction, with the refracted ray traversing the glass at a certain angle of refraction;

(c) it is absorbed by the glass, so that its intensity is reduced and its spectral distribution changed in accordance with the absorptive capacities of the glass;

(d) it is scattered by the scattering centres present in the glass;

(e) it is doubly-refracted, depending upon the degree of isotropy of the glass;

(f) it is polarized due to reflection, refraction, absorption, scattering, etc.;

(g) interference occurs.

Figure 30 is a diagram indicating the principal optical effects characteristic of the interaction of radiation with glass, viz. reflection, refraction, absorption, and scatter.

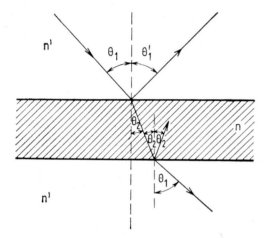

Fig. 30 Reflection and transmission at a boundary.

The incident radiant flux, Φ, will thus, in the simplest case, split into beams of reflected, refracted, transmitted, absorbed, and scattered radiation, so that we can write

$$\Phi = \Phi_{\varrho r} + \Phi_{\varrho d} + \Phi_{\tau r} + \Phi_{\tau d} + \Phi_{\alpha}, \tag{73}$$

where Φ is the total incident flux,

$\Phi_{\varrho r}$ the directly reflected flux,

$\Phi_{\varrho d}$ the diffusely reflected flux,

$\Phi_{\tau r}$ the transmitted flux,

$\Phi_{\tau d}$ the scattered flux

and Φ_{α} the absorbed flux.

Then

$$1 = \frac{\Phi_{\varrho r}}{\Phi} + \frac{\Phi_{\varrho d}}{\Phi} + \frac{\Phi_{\tau r}}{\Phi} + \frac{\Phi_{\tau d}}{\Phi} + \frac{\Phi_{\alpha}}{\Phi}, \tag{74}$$

where $\dfrac{\Phi_{\varrho r}}{\Phi}$ is the coefficient of specular reflection, ϱ_r,

$\dfrac{\Phi_{\varrho d}}{\Phi}$ the coefficient of diffuse reflection, ϱ_d,

$\dfrac{\Phi_{\tau r}}{\Phi}$ the coefficient of direct transmission, τ_r,

$\dfrac{\Phi_{\tau d}}{\Phi}$ the coefficient of scattered transmission, τ_d,

and $\dfrac{\Phi_{\alpha}}{\Phi} = \alpha$ the absorptance (absorption factor).

Then

$$(\varrho_r + \varrho_d) + (\tau_r + \tau_d) + \alpha = 1, \tag{75}$$

and

$$\varrho + \tau + \alpha = 1, \tag{76}$$

where ϱ is the coefficient of reflection (total reflectance),

τ the coefficient of transmission (total transmittance),

and α the coefficient of absorption (absorptance).

Birefringence, polarization, and other effects associated with the interaction of incident radiant flux with glass are not represented in Figure 30, neither are they considered in equation (73).

In certain special glasses radiation gives rise to some quite exceptional physico-chemical effects, e.g., spontaneous emission, stimulated emission, photochemical effects, photoelectric effects, etc. These glasses, which possess special optical properties, will be dealt with in more detail in later chapters and sections of this book.

In concluding this chapter, it needs to be emphasized that the character of the radiation and the associated optical effects should be studied from the points of view of three basic disciplines:

(a) from the point of view of geometrical optics,

(b) from the point of view of wave optics, and

(c) from the point of view of quantum optics.

Refererces

[1] M. Fanderlik, Struktura skel (Structure of Glasses). SNTL, Prague 1971.
[2] M. Garbuny, Optical Physics. Academic Press, New York—London 1965.
[3] V. Prosser, Optické vlastnosti pevných látek (Optical Properties of Solids) (Mimeographed). SPN, Prague 1971.
[4] V. Vrba, Moderní aspekty klasické fyzikální optiky (Modern Aspects of Classical Optical Physics). Academia, Prague 1974.

[5] D. G. Holloway, The Physical Properties of Glass. Wykeham Publ. Ltd., London—Winchester 1973.

[6] Kolektiv autorů, Teorie pevných látek (Collection of Contributions by Various Authors, Theory of Solids). NČSV, Prague 1965.

[7] J. D. Mackenzie, Modern Aspects of the Vitreous State. Vols. I, II, III. Butterworth, London 1960—1964.

[8] H. Rawson, Inorganic Glass-Forming Systems. Acad. Press, London 1967.

[9] W. A. Weyl and E. Ch. Marboe, The Constitution of Glasses. A Dynamic Interpretation. Vols. I and II. Wiley, New York 1962, 1964, 1966.

[10] Vl. Šatava, Povaha skelného stavu a podmínky tvorby skla (The Nature of the Vitreous State and the Glass-Forming Conditions). Československý časopis pro fyziku 23 (1973) No. 6, pp. 565—583.

[11] D. Turnbull and M. H. Cohen, Crystallization Kinetics and Glass Formation. In: Modern Aspects of the Vitreous State. Vol. I. Butterworths, London 1960.

[12] D. Turnbull, Solid State Physics. Vol. 3. Academic Press. New York 1956.

[13] W. D. Hillig, in: Symposium on Nucleation and Crystallization in Glasses and Melts. Amer. Ceram. Soc., Columbus, Ohio, 1962.

[14] D. R. Uhlmann, Amorphous Materials. Wiley-Interscience, London 1972.

[15] W. Vogel, Glaschemie (Glass Chemistry). VEB Deutscher Verlag für Grundstoffindustrie, Leipzig 1979.

[16] H. Scholze, Glasnatur, Struktur und Eigenschaften. Springer Verlag, Berlin—Heidelberg—New York 1977.

[17] J. W. Cahn, Phase separation by spinodal decomposition in isotropic systems. J. Chem. Phys., 42 (1965) pp. 93—99.

[18] J. W. Cahn and J. R. Charles, The initial stages of phase separation in glasses. Phys. Chem. Glasses 6 (1965) pp. 181—191.

[19] J. W. Cahn and J. E. Hilliard, Free energy of non-uniform systems. III. Nucleation in a Two-component incompressible fluid. J. Chem. Phys., 31 (1959) pp. 688—699.

[20] W. Haller, Rearrangement kinetics of the liquid-liquid immiscible microphases in alkaliborosilicate melts. J. Chem. Phys. 42 (1965) No. 2, pp. 686—693.

[21] W. Haller, H. D. Blackburn, F. E. Wagstaff, and R. J. Charles, Metastable immiscibility surface in the system Na_2O—B_2O_3—SiO_2. J. Amer. Cer. Soc. 53 (1970)No. 1, pp. 34—39.

[22] J. Zarzyckij, Phase-separated systems. Discussions Farady Soc. 50 (1970) pp. 122—144.

[23] C. L. Babcock, Silicate Glass Technology Methods. Willey, New York 1977.

[24] J. Wong and C. A. Angell, Glass Structure by Spectroscopy. M. Dekker, New York 1976.

[25] P. Balta, Introduction of the Physical Chemistry of the Vitreous State. Academia—Abacus Press, Tunbridge Wells—Kent 1976.

[26] A. M. Alper, Phase Diagrams. Academic Press, New York—London 1970.

[27] J. M. Ciwley, Diffraction Physics. North-Holland Publ. Co., Amsterdam 1975.

[28] N. S. Andreyev, O. V. Mazurin, E. A. Poray—Koshits, E. A. Roskova, and V. N. Filipovitch, Iavlenya likvatsyi v steklakh (Effects of Liquation in Glasses). Nauka, Leningrad 1974.

[29] J. Frenkel, The Kinetic Theory of Liquids. Clarendon Press, Oxford 1946.

[30] W. Vogel, Struktur und Kristallisation der Gläser. VEB Deutscher Verlag für Grundstoffindustrie, Leipzig 1965.

[31] J. M. Stevels, Progress in the Theory of the Physical Properties of Glass. Elsevier, Amsterdam 1944.

[32] J. E. Stanworth, Physical Properties of Glass. Clarendon Press, Oxford 1950.

[33] J. A. Prins, Physics of Non-Crystalline Solids. North-Holland, Amsterdam 1965.

[34] N. F. Mott and E. A. Davis, Electronic Processes in Non-Crystalline Materials. Clarendon Press, Oxford 1971.

[35] I. Fanderlik, Optické vlastnosti skel (Optical Properties of Glasses). SNTL, Prague 1979.
[36] M. Tomozawa, R. H. Doremus, Treatise on Materials Science and Bechnology. Vol. 17, Glass II: Phase Separation in Glass (M. Tomozawa). Academic Press, New York, London, Toronto, Sydney, San Franciscn, 1979 pp. 71—113.

4.1   REFLECTION OF RADIATION BY GLASSES

4.1.1   Introduction

As already mentioned in Chapter 4, part of the radiation incident upon the surface of glass is reflected, part is refracted, absorbed, scattered, etc. The magnitude and the character of the reflected radiation depends upon the quality of the interface, upon the angle of incidence, upon the difference between the refractive indices of the glass and the medium from which the radiation emerges (or vice versa), and upon the wavelength of the radiation.

In glasses that are highly absorbing in the visible region of the spectrum (or in the infra-red region) the remaining optical parameters can be computed from the measurements of the changes in the parameters of the radiation reflected by an optically defined surface.

4.1.2   Theory

If an incident beam of monochromatic radiation proceeding from a medium the refractive index of which is n' (e.g. air) reaches the interface separating that medium from a medium whose refractive index is n (glass), it is in general split into two rays (see Figure 30). One ray remains in the n'-indexed medium: this is the reflected ray; the other is refracted into the medium with the n index of refraction, where as we will explain below, it either passes through direct, or is absorbed, scattered, etc.

The angle which is included between the incident ray and the line normal to the interface at the point of incidence, Θ_1, is the angle of incidence; Θ_2 is the angle of refraction, and Θ'_1 the angle of reflection. The direcion of the refracted or reflected ray is given by Snell's law of refraction, viz.

$$n \sin \Theta_2 = n' \sin \Theta_1, \tag{77}$$

or reflection

$$\Theta'_1 = \Theta_1. \tag{78}$$

If the ray passes from an optically rarer medium into an optically denser medium, viz. $n' < n$, then $\Theta_2 < \Theta_1$, and the ray is refracted towards the normal. On the other hand, if the ray passes from an optically denser into an optically rarer medium, viz. $n' > n$, then $\Theta_2 > \Theta_1$, an the refraction is away from the normal. In the latter case angles Θ_1, Θ'_1, and Θ_m are equal for a certain angle of incidence, Θ_m being the marginal, or critical, angle. For angles of incidence larger than the critical angle Θ_m the whole radiation is reflected by the interface. This is the case of the so-called *full*, or *total, reflection*.

If a monochromatic ray of light emerging from medium n' meets the surface of glass n perpendicularly, its coefficient of reflection (reflectance), ϱ, on one surface can be calculated from Fresnell's formula, viz.

$$\varrho = \left(\frac{n - n'}{n + n'}\right)^2,$$

(79)

where ϱ is the coefficient of reflection (reflectance) on one surface,
$\quad\quad n'$ the refractive index of incident medium
and $\quad n$ the refractive index of glass.

Thus, for instance, for glass of refractive index 1.5, we obtain a coefficient of reflection on one surface, ϱ, equalling 0.04 (4 %), for glass of refractive index 1.75 reflectance ϱ will be 0.075 (7.5 %), and for glass with index of refraction $n = 1.9$, the coefficient of reflection ϱ, will equal 0.097 (9.7 %).

Table 9

n	Loss of light due to reflection $\varrho \cdot 100$ [%]							
	1 surface	2 surfaces	4 surfaces	6 surfaces	8 surfaces	10 surfaces	12 surfaces	14 surfaces
1.45	3.37	6.63	12.82	18.61	24.00	29.00	33.75	38.14
1.50	4.00	7.84	15.07	21.72	27.66	33.52	38.73	43.53
1.55	4.65	9.09	17.35	24.86	31.69	37.90	43.54	48.67
1.60	5.33	10.37	19.66	27.99	35.45	42.14	48.14	53.52
1.65	6.01	11.67	21.98	31.08	39.13	46.13	52.51	58.05
1.70	6.72	12.99	24.29	34.12	42.68	50.13	56.61	62.24
1.75	7.44	14.32	26.60	37.11	46.12	53.84	60.45	66.11
1.80	8.16	15.66	28.87	40.01	49.40	57.33	64.01	69.65

For normal incidence and for a larger number of surfaces, with absorptivity equalling zero, reflectance ϱ can, according to [4], be computed from the relationship

$$\varrho = 1 - \left[1 - \left(\frac{n - 1}{n + 1}\right)^2\right]^{m'},$$

(80)

where m' is the number of boundary surfaces.

In Table 9 the reader will find the coefficients of reflection of glasses of different refractive indices and of different numbers of boundary surfaces calculated from equation (80) for zero absorptivity.

From Table 9 it follows that, for instance, in sophisticated optical instruments with twelve boundary surfaces the loss of light due to reflection can amount to as much as 50 %, dependent, of course, upon the refractive index of the glass. It is therefore most advisable that antireflection coatings should be applied in these cases so as to reduce losses due to reflection. The subject will be dealt with more comprehensively in Chapter 7.

Metals exhibit a generally high reflectance, because they contain free electrons. The frequency of the incident radiation thus resonates with the frequencies of these free electrons, which gives rise to absorption. The major part of the incident radiation is reflected, the minor part is converted into heat. In optical practice, use is often made of silver, gold, and aluminium as materials for reflecting surfaces, for they manifest high reflectivity, as much as 99 %.

For conducting materials like silver, gold, and aluminium reflectance can be computed by developing equation (88) in [2], assuming that $\mu \doteq 1$,

$$n = k' = \sqrt{\frac{\sigma}{\nu}}, \quad \frac{\sigma}{\nu} \geq 1:$$

$$\varrho = \frac{2\frac{\sigma}{\nu} - 2\sqrt{\frac{\sigma}{\nu} + 1}}{2\frac{\sigma}{\nu} + 2\sqrt{\frac{\sigma}{\nu} + 1}} = 1 - 2\sqrt{\frac{\nu}{\sigma}} + ..., \tag{81}$$

where σ denotes electrical conductivity,
and ν frequency.

Reflectance can also be calculated from equation (88) quoted in Section 4.1 [2] by substituting into it the numerical values for n^2 and k'^2 obtained from the following equations:

$$n^2 = \frac{\mu}{2}\left[\sqrt{\varepsilon^2 + \left(\frac{4\pi\sigma}{\omega}\right)^2} + \varepsilon\right], \tag{82}$$

$$k'^2 = \frac{\mu}{2}\left[\sqrt{\varepsilon^2 + \left(\frac{4\pi\sigma}{\omega}\right)^2} - \varepsilon\right], \tag{83}$$

where μ denotes permeability,
ε permittivity,
σ electrical conductivity,
ω angular frequency.

Table 10 lists the results of the measurements and calculations of the coefficient of reflection from equations (81), (82), and (83) for silver, gold and aluminium (after [2]).

So far we have assumed that the surfaces reflecting the light are optically smooth planar surfaces. If, however, the surface is rough or nonplanar (i.e., every part of the surface has different orientation), the incident radiation is reflected, or refracted,

Table 10

Metal	λ [μm]	n	k'	$\varrho_1 \cdot 100$ [%]	$\varrho_2 \cdot 100$ [%]	$\varrho_3 \cdot 100$ [%]
Al	0.492	0.68	4.80	90 (MV)		91.2
	0.950	1.75	8.50		90 (ae)	93.6
	2.000	2.30	16.50	98.4 (MV)	95 (ae)	95.6
	5.000	33.60	76.40	99.1 (MV)	98 (p)	98.2
	12.000	8.19	36.80	98.8 (MV)	96 (p)	97.3
Ag	0.226	1.406	1.107	19.8 (p)	20	
	0.316	1.127	0.427	4.2 (p)	5	
	0.492	0.123	2.720	94.3 (ae)		93.3
	1.000	0.240	6.960	98.1 (ch)	97.0 (p)	95.3
	2.000	0.680	13.7	98.5 (ch)	97.8 (p)	96.7
	4.370	4.34	32.6	98.6 (ch)	98.5 (p)	97.7
	10.000				99.0	98.5
Au	0.508	0.908	2.075	54.3 (ea)	50 (p)	92.0
	1.000	0.194	5.570	97.5 (ae)		94.3
	2.000	0.560	15.400	99.1 (ae)	96.5 (ad)	96.0
	3.140	0.800	18.900	99.2 (ea)		96.8
	4.830	1.830	33.000	99.3 (ea)	97.2	97.4
	9.000				98.0	98.1

$\varrho_1 \cdot 100$ — calculated from measurement of n and k' (equations (82) (83)),
$\varrho_2 \cdot 100$ — measured,
$\varrho_3 \cdot 100$ — calculated from equation (81) with values of σ for direct current, and $\mu = 1$,
MV — measured in vacuo,
ae — applied by evaporation,
p — polished,
ch — chemically applied,
ea — electrolytically applied,
ad — applied by dusting.

in each of these surfaces in a different direction and the beam of parallel rays is scattered. A rough surface produces, apart from the specular reflection already mentioned above, diffuse reflection [5, 6]. If the surface of the glass manifests irregularities the orientations of which are evenly distributed, all directions of reflection are equally represented in the reflected light, with no direction preferred. The response of a surface of this type to incident radiation is termed diffusely scattered reflection. Typical examples of diffusely scattering surfaces are layers of newly precipitated barium sulphate, calcium sulphate, or magnesium oxide.

Figure 31 gives a schematic representation of the angular distributions of the intensity of reflected radiation for specular, mixed and diffuse reflection in the form of polar diagrams.

The reflected ray is *polarized* and it oscillates in the plane perpendicular to the plane of incidence. The degree of the polarization depends upon the angle of incidence, Θ_1. It can be totally polarized, if the angle of incidence corresponds to the so-called *Brewster's angle of polarization*, Θ_p. These problems are dealt with in more detail in Section 4.5.

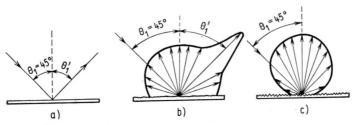

Fig. 31 Schematic representation of specular (a), mixed (b), and diffuse (c) reflection.

Definition of the reflection of radiation

The effect of the radiation returning from the interface back into its parent medium, with the frequency of its monochromatic components remaining the same, is called *reflection*. For all the angles of incidence, Θ_1, of the radiation at the interface between an opticaly denser and an optically rarer medium $(n < n')$ which are larger than the critical angle, Θ_m, the reflection is complete and total. In that case $\Theta_m = \Theta'_1$.

The unscattered reflection of light obeying the laws of optical reflection is called regular, or specular, reflection. The reflection of light in various directions, with no regular reflection produced in the macroscopic scale, is called diffuse reflection. The term mixed reflection applies if a surface produces both specular and diffuse reflection simultaneously.

If the diffuse reflection is uniform, the reflecting surface will in all directions manifest the same radiance and luminance.

The coefficient of reflection (reflectance), ϱ, is defined as the ratio of the radiant (luminous) flux reflected to the incident radiant (luminous) flux. Thus,

$$\varrho = \frac{\Phi_{\varrho r} + \Phi_{\varrho d}}{\Phi} . \tag{84}$$

4.1.3   Dependence of reflectance upon the angle of incidence of monochromatic radiation

If the monochromatic radiation travelling in air $(n' = 1)$ meets the surface of glass the refractive index of which is $n = 1.70517$, the curve for the dependence of the reflectance upon the angle of incidence, Θ_1, will, according to [7], be as shown in Figure 32.

On the other hand, if monochromatic radiation travelling in glass ($n = 1.70517$) meets the glass-air boundary, the curve for the dependence of the coefficient of reflection, ϱ, upon the angle of incidence, Θ_1, will be of a slightly different shape (see Fig. 33).

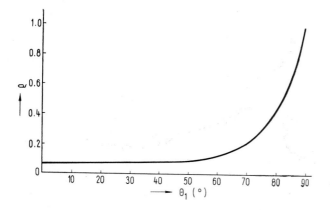

Fig. 32 Dependence of reflectance, ϱ, upon the angle of incidence, Θ_1, for monochromatic radiation incident in the less dense medium.

In this case the critical angle is $\Theta_m = 36°30'$. The coefficient of reflection shows little dependence upon the angle of incidence, Θ_1, of monochromatic radiation for the angles up to about 50° in the former case (Fig. 32), and about 30° in the latter case (Fig. 33), equalling approximately the reflectance for perpendicular incidence, viz. for the angle 0°. This fact is taken an advantage of in technological practice if calculations are to be simplified.

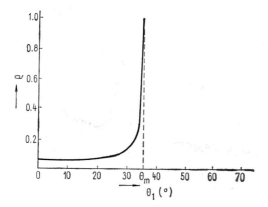

Fig. 33 Dependence of reflectance, ϱ, upon the angle of incidence, Θ_1, for monochromatic radiation incident in the dense medium.

As already mentioned above, the critical angle, Θ_m, depends upon the index of refraction of the glass. In Figure 34 the reader will find this dependence graphically represented.

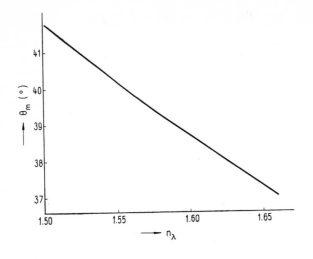

Fig. 34 Dependence of critical angle, Θ_m, upon the refractive index of glass, n_λ.

4.1.4 Dependence of reflectance upon the wavelength of monochromatic radiation

In high accuracy computations the dependence of the coefficient of reflection upon the wavelength of monochromatic radiation must be taken due account of. (See Figure 35). Even though this dependence may not seem to be of

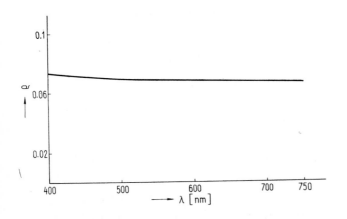

Fig. 35 Dependence of reflectance, ϱ, at normal incidence, upon the wavelength, λ, of incident radiation.

decisive importance for calculations within the range of wavelengths in the visible region of the spectrum, it will nevertheless have to be paid attention to in certain cases.

As compared with the transparent, non-absorbing glasses, the dependence of the coefficient of reflection upon the wavelength is, however, fairly marked in coated glasses and metals.

4.1.5 Dependence of reflectance upon the refractive index of the glass

It follows from Fresnell's formula (79) that the coefficient of reflection (reflectance), ϱ, depends upon the refractive index of the glass, with reflectance being considered in one reflecting surface only in the case quoted (see Table 11). For a larger number of boundary surfaces the coefficient of reflection can be computed from equation (80). The relationships have been derived for non-absorbing media, including colourless glasses. For the computation of the reflection of monochromatic radiation traversing a plate of glass (reflection on two surfaces) use can be made in this case of the so-called factor R, reflection factor, which can be calculated from simple relationships, viz.

$$R = \frac{2n}{n^2 + 1} \,, \tag{85}$$

or

$$R = 1.287 - 0.243\, n. \tag{86}$$

Table 11

n	ϱ	R	n	ϱ	R
1.45	0.034	0.935	1.59	0.052	0.901
1.46	0.035	0.932	1.60	0.053	0.898
1.47	0.036	0.930	1.61	0.055	0.896
1.48	0.038	0.927	1.62	0.056	0.894
1.50	0.040	0.923	1.63	0.057	0.891
1.51	0.041	0.920	1.64	0.059	0.888
1.52	0.043	0.918	1.65	0.060	0.886
1.53	0.044	0.915	1,66	0.062	0.884
1.54	0.045	0.913	1.67	0.063	0.881
1.55	0.047	0.910	1.68	0.064	0.879
1.56	0.048	0.908	1.69	0.066	0.876
1.57	0.049	0.905	1.70	0.067	0.874
1.58	0.051	0.903	1.80	0.082	0.850
			1.90	0.096	0.825

For the refractive index of glass, n, we invariably substitute values recorded for $\lambda_d = 587.56$ nm or $\lambda_e = 546.07$ nm, or $\lambda_D = 589.29$ nm. The results of calculations based on equations (85) and (86) are accurate enough to be applicable in practice. Table 11 tabulates the values of R for glasses of various indices of refraction together with the values of ϱ. Factors R (reflection factors) are mostly applied in the conversion of the recorded values of the propagation of radiation through glass into the values of the uninhibited transit free from losses due to reflection.

The more complex and sophisticated computations of reflectance are dealt with in Chapter 7.

4.1.6 Dependence of reflectance upon the chemical composition of the glass

The refractive index of glass depends upon its chemical composition. This fact has already been emphasized and will again be dealt with in Section 4.2. It follows that the coefficient of reflection will also be dependent upon the chemical composition of the glass, which, in this case, is indirectly expressed by equations (79) and (80).

4.1.7 Dependence of reflectance upon the absorptive properties of the glass

The reflectance of absorbing glasses does not depend solely upon their indices of refraction, but upon their *coefficient of absorption, a,* as well. According to [2, 15] we get a simpler expression (than equation (81)) for the calculation of the reflectance, ϱ, if the *absorption factor, k',* is computed from

$$k' = \frac{a}{4\pi} \lambda_0, \tag{87}$$

where $\lambda_0 = \lambda_n$,
and substituted into equations (88) or (89).

Thus, if the radiation leaving the air medium meets the boundary between air and the glass (absorptive glass) at normal incidence we get for the reflectance the following relationship:

$$\varrho = \frac{(n - \mu)^2 + k'^2}{(n + \mu)^2 + k'^2}. \tag{88}$$

Since for the dielectrics it holds approximately that $\mu \doteq 1$, equation (88) can be rewritten as

$$\varrho = \frac{(n - 1)^2 + k'^2}{(n + 1)^2 + k'^2}. \tag{89}$$

In equations (87), (88), and (89)
a denotes the linear absorption coefficient,
n the refractive index of the glass,
k' the absorptive factor,
μ permeability,
λ the wavelength.

If the absorption of the radiant flux by glass is negligible, then $k' = 0$, and equation (89) will correspond to Fresnell's formula (79).

4.1.8 Measuring methods and apparatus

Introduction

As already mentioned above, the reflection of radiation depends, among other things, upon the quality, or the character, of the surface of the glass, the

72

measure thus being the ratio of the reflected to the incident radiant flux. And the reflection depends not only upon the wavelength and the polarization of the radiation but also upon the angle of incidence. For measuring purposes the angle of incidence has been standardized, viz. $\Theta_1 = 45°$, according to [5]. Reflection is thus given as a value related to a definite and standard surface, most often to a surface coated by a newly precipitated layer of barium sulphate, calcium sulphate, or magnesium oxide, or to a surface of silver-, or aluminium-, plated mirrors.

And these facts also influence our decisions concerning the different designs of the measuring devices, viz. reflectometers or glossmeters. Both visual and photoelectric reflectometers exist. Some of these instruments are designed in such a way that they can record radiation reflected at an angle of 45° only, for others the range of application may be far larger.

Photoelectric reflectometers

Measuring reflection by the Pirani-Schönborn method

This method of measurement makes it possible to record the reflection of a perpendicularly incident luminous flux [8]. Use is made of a small integrating sphere, with the luminous flux emitted by source S concentrated on to the specimen of the glass V to be examined by a condenser. The plate coated with a uniformly scattering material (magnesium oxide) is subjected to measurement first, the same

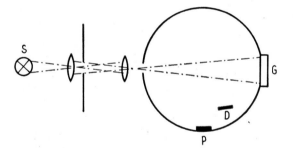

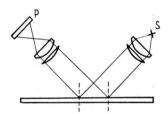

Fig. 36 Diagram of apparatus measuring the reflection of light by the Pirani-Schönborn method.

Fig. 37 Diagram of apparatus measuring the reflection of light with the help of the Lange glossmeter.

coating being applied to the inner surface of the integrating sphere. The luminous flux is diffusely reflected over the whole inner surface of the sphere, and a photoelectric sensor, P, records the signal (Fig. 36). As the next step, the glass specimen, G, is inserted and the measuring procedure is repeated; the value obtained is the diffusely reflected luminous flux.

If measurement is to be made of mixed reflection (specular together with diffuse), the glass specimen, G, is deflected a bit from its position normal to the direction of the incident luminous flux, and the reading can be taken of the photoelectric sensor, *P. D* is a diaphragm.

The results of the measurement are then related to the coefficient of reflection (reflectance) of the plate coated with magnesium oxide.

Lange's glossmeter

Lange's device records the reflection of luminous flux at an angle of 45°. A schematic diagram of the apparatus is shown in Figure 37. A silver-coated mirror the reflectance of which is about 99 per cent is used as the comparison surface.

Measurement made on the mirror is followed by measurement on the glass specimen. The results are then related to the reflectance of the silver-coated mirror. If the surface of the glass does not exhibit perfect gloss, we have the case of mixed reflection.

Spekol spectrophotometer with R 45/0 adapter

This apparatus is a grating-type monochromator equipped with an R 45/0 adapter measuring the reflected monochromatic luminous flux at an angle of 45°. A schematic representation of the apparatus can be seen in Figure 38. The adapter is designed so that the luminous flux emerging from the slit of the monochromator, *A*, meets the glass specimen, *G*, at an angle of 45°, after being reflected by mirror M_1. The luminous flux reflected normal to the specimen passes through lenses L_2 and L_1 on to the photoelectric sensor. (*B* is a slit.) The results of the measurement are then compared with those for the newly precipitated layer of barium sulphate or magnesium oxide.

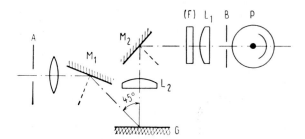

Fig. 38 Diagram of apparatus measuring the reflection of light the Specol spectral photocolorimeter with an R 45/0 adapter.

Since the R 45/0 adapter is attached to a grating-type monochromator, reflection can be measured for the various wavelengths of the visible region of the spectrum. Fluorescence, which may appear in some cases, can be removed by interference filters, *F*, inserted into this apparatus, which let through incident radiant flux only.

Measuring reflection of approximately perpendicular incident luminous flux

Figure 39 gives a schematic diagram of the device. The light emitted by source *S* passes through monochromator *MC*, or polarizer *PL*, on to a photoelectric sensor *PH* (with the specimen under examination, *G*, removed); the deviation of the galvanometer is set for 100 % (Fig. 39a). After the insertion of the specimen the luminous flux will be reflected by the surface of the glass specimen under examination. With this arrangement the angle of incidence Θ_1 does not equal zero, but for angles $\Theta_1 < 5°$ the error caused by oblique incidence is negligible. The accuracy of the measurement is reduced by the necessity of moving the specimen and the photoelectric sensor (see Figure 39a).

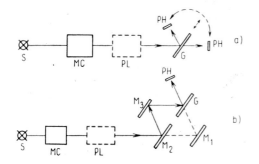

Fig. 39 Schematic diagram of apparatus measuring the reflection of light at normal incidence:

a — with an ancillary mirror; b — with three ancillary mirrors.

A method has therefore been devised making use of three auxiliary mirrors, M_1, M_2, and M_3, which must be metal-coated in one batch (for the coating must exhibit equal quality). (See Fig. 39b.) Mirrors M_1 and M_3 are fixed in their positions, mirror M_2 is mounted on a movable support together with the specimen of the glass under examination, G. The luminous flux emerging from monochromator MC is reflected before contacting photoelectric sensor PH:

(a) by mirror M_1 only — luminous flux Φ_1,

(b) successively by specimen G and mirrors M_3 and M_2 — luminous flux Φ_2,

(c) successively by mirrors M_2, M_3, and again M_2 — luminous flux Φ_3.

Assuming that mirrors M_1, M_2, and M_3 manifest equal reflectance, we can calculate the reflection coefficient of the glass specimen under examination from the following relationship:

$$\varrho = \Phi_2(\Phi_1\Phi_3)^{-1/2}. \tag{90}$$

Cary 14 recording spectrophotometer

For reflectance and transmittance measurements this apparatus can be equipped with an 50-601-010 universal adapter.

4.1.9   Calculations

In glassworks practice the reflection of luminous flux is paid particular attention to in calculations of spectral transmission, viz. in the conversion of the values of spectral transmission recorded into those of the transmission free from reflection-incurred losses. In other cases we are interested in the intensity of the reflected radiant flux.

Some simple examples are given below.

Example 1

Calculate the percentage of the monochromatic flux reflected by one surface of glass, if the refractive index of the glass, n_d, equals 1.52, and if the monochromatic flux leaving the air medium meets the surface of the glass at angle of normal incidence. Applying equation (79), we get

$$\varrho \cdot 100 = \left(\frac{n-1}{n+1}\right)^2 \cdot 100 = \left(\frac{1.52-1}{1.52+1}\right)^2 \cdot 100 = 4.3\%. \tag{91}$$

Example 2

Calculate the percentage of the monochromatic flux reflected by two surfaces of glass, if the refractive index of the glass $n_d = 1.52$, the other conditions being the same as those in Example 1.

Table 12

λ [nm]	$\tau \cdot 100$ [%] recorded	$\tau_t \cdot 100$ [%] free from losses due to reflection	λ [nm]	$\tau \cdot 100$ [%] recorded	$\tau_t \cdot 100$ [%] free from losses due to reflection
340	12.5	13.6	540	61.0	66.4
360	55.0	59.8	560	60.5	65.8
380	56.5	61.5	580	59.5	64.7
400	69.5	75.6	600	56.0	60.9
420	48.0	52.2	620	49.0	53.3
440	38.5	41.9	640	47.0	51.1
460	40.5	44.1	660	48.5	52.8
480	46.0	50.1	680	54.5	59.3
500	54.0	58.8	700	55.0	59.7
520	60.0	65.3			

Applying equation (80), we get

$$\varrho \cdot 100 = \left\{1 - \left[1 - \left(\frac{n-1}{n+1}\right)^2\right]^{m'}\right\} \cdot 100 =$$

$$= \left\{1 - \left[1 - \left(\frac{1.52-1}{1.52+1}\right)^2\right]^2\right\} \cdot 100 = 8.3\%; \tag{92}$$

and for four surfaces we get

$$\varrho \cdot 100 = \left\{1 - \left[1 - \left(\frac{1.52-1}{1.52+1}\right)^2\right]^4\right\} \cdot 100 = 16\%. \tag{93}$$

Example 3

Convert the recorded values of spectral transmittance τ_λ of Greenal sun glass into the values free from reflection losses, if the refractive index of the glass, n_d, is equal to 1.518. With the help of the values tabulated in Table 11 defining the dependence of reflection factor **R** upon the refractive index of glass, or by way of a calculation from equations (85) and (86), we get for the given refractive index the reflection factor **R** = 0.919. Dividing the recorded percentile values of spectral transmission factor τ by factor **R**, we obtain the values of the spectral transmission factor free from reflection losses, τ_i (see Table 12).

References

[1] B. Havelka, E. Keprt, and M. Hansa, Spektrální analýza (Spectral Analysis). NČAV, Prague 1957.
[2] M. Garbuny, Optical Physics. Academic Press, New York—London 1965.
[3] V. Prosser, Optické vlastnosti pevných látek (Optical Properties of Solids). SPN, Prague 1971.
[4] J. Hajda, Optika a optické přístroje (Optics and Optical Instruments). SVTL, Bratislava 1956.
[5] Committee on Colorimetry—Optical Society of America, The Science of Color. T. Y. Crowell Co., New York 1953.
[6] J. Velíšek, Povaha lesku a metody jeho měření (Nature of Gloss and Methods of Measurement of Gloss). *Sklářské rozhledy, 20* (1943) No 1, pp. 4—12.
[7] B. Havelka, Geometrická optika (Geometrical Optics). NČAV, Prague 1955 (Vol. I) 1956 (Vol. II).
[8] M. Fanderlik, Sklářské praktikum (Glass Makers Practicum). Průmyslové vydavatelství, Prague 1951.
[9] J. Götz et al., Broušení a leštění skla (Glass Cutting and Grinding). SNTL, Prague 1963.
[10] Kolektiv autorů (Collective of Authors), Osvětlovací sklo v interiéru (Interior Lighting Glass). SNTL, Prague 1965.
[11] Zd. Horák, Fr. Krupka, and V. Šindelář, Technická Fyzika (Technical Physics). SNTL, Prague 1961.
[12] J. Hajda, Technická optika (Technical Optics). Práce, Prague 1951.
[13] V. Petržílka, Fyzikální optika (Optical Physics). Přírodovědecké nakladatelství, Prague 1952.
[14] M. Born and E. Wolf, Principles of Optics. Pergamon Press, New York 1964.
[15] A. P. Prishivalko, Otrazhenyie sveta ot pogloshchayushchikh sred (Reflection of Light Radiation by Absorbing Media). Minsk 1963.
[16] A. V. Sokolov, Opticheskye svoystva metalov (Optical Properties of Metals). GIFML, Moscow 1961.
[17] Ellipsometry in the Measurement of Surfaces and Thin Films. Symposium Proceedings, Washington 1963, NBS Miscellaneous Publ. 256, 1964.
[18] W. M. Wendlandt and H. G. Hecht, Reflectance Spectroscopy. Interscience Publishers, New York 1966.
[19] G. Kortüm, Reflexionsspetroskopie. Grundlagen, Methodik, Anwendungen (Reflection Spectroscopy, Foundations, Methods, Applications). Springer Verlag, Berlin 1969.
[20] H. Y. Fan, in: Methods of Experimental Physics (L. Marton ed. Vol 6, Part B, p. 252). Academic Press, New York 1959.
[21] J. Brož et al., Základy fyzikálních měření (II B) (Foundations of Physical Measurement). SPN, Prague 1974.

4.2 REFRACTION AND DISPERSION OF RADIATION BY GLASSES

4.2.1 Introduction

In Section 4.1.1 we stated that the refraction of a ray of light by glass is defined by Snell's law, which, in case $n' = 1$, can be rewritten as

$$n(\lambda) = \frac{\sin \Theta_1}{\sin \Theta_2} = \frac{c}{v_\lambda}, \qquad (94)$$

where c is the velocity of light in vacuo,
and v_λ the phase velocity of light in glass.

The velocity of a ray of light passing through a material medium (glass), and hence the corresponding refractive index, n, also, depend upon the interaction of the electric field associated with the ray and the outer electron shells of the atoms which constitute the material of the medium. The luminous energy makes the electrons oscillate harmonically, so that the outer electrons can be viewed as harmonic oscillators.

The phase velocity of the light, v_λ, will then decrease in proportion to the polarizability of the outer electrons. When the electrons and the frequency of the light are in perfect resonance, we have the case of the absorption. These problems are dealt with in Section 4.3.

4.2.2 Theory

If we suppose every electron of charge e and mass m to be bound in its equilibrium position by an elastic force, the movement of that electron, in the absence of an external field, can be defined by the equation of harmonic motion [1, 2]

$$m\ddot{x} = -kx, \qquad (95)$$

where $\ddot{x}$ is the acceleration,
$\quad$ k equals $m\omega^2$ (where ω denotes angular frequency),
and $\quad$ x is the displacement from the equilibrium position.

Equation (95) has the following solution:

$$x = x_0 \cos 2\pi v_0 t, \qquad (96)$$

where v_0 is the natural frequency of the vibration of the electrons, viz.

$$v_0 = \frac{1}{2\pi} \sqrt{\frac{k}{m}}. \qquad (97)$$

In the presence of an electric field, the electron is subject to a force E, which changes periodically with frequency v. Then, if $E = E_0 \cos 2\pi v t$, we can write

$$m\ddot{x} = -kx + eE_0 \cos 2\pi vt. \tag{98}$$

For the large values of t, the solution is

$$x = x_0 \cos 2\pi vt. \tag{99}$$

Substituting this solution into equation (98), we get

$$-m\,4\pi^2 v^2 x_0 \cos 2\pi vt = -kx_0 \cos 2\pi vt + eE_0 \cos 2\pi vt. \tag{100}$$

After reduction, we have the following relationships for the amplitude of the vibrations of the electron:

$$x_0 = \frac{e}{k - 4\pi^2 mv^2} E_0, \tag{101}$$

$$x = \frac{e}{k - 4\pi^2 mv^2} E_0 \cos 2\pi vt. \tag{102}$$

The displacement of the electrons owing to the effect of the light wave is thus defined. Multiplying by e, the charge of the electrons, we get the expression for the induced dipole moment

$$p = ex = \frac{e^2}{k - 4\pi^2 mv^2} E = \alpha E. \tag{103}$$

From equation (103), by substituting k computed from equation (97), we obtain

$$\alpha = \frac{e^2}{4\pi^2 m}\frac{1}{v_0^2 - v^2} = \frac{e^2}{m\omega_0^2 - \omega^2}, \tag{104}$$

where $\omega_0 = 2\pi v_0$, and
$$\omega = 2\pi v.$$

Equation (104) gives the coefficient of optical polarizability of one electron as a function of the frequency of the incident light v. For the coefficient of static polarizability, α_0, with the external field being constant: $\lambda \to \infty$; $v \to 0$, we can write

$$\alpha_0 = \frac{e^2}{4\pi^2 mv_0^2} = \frac{e^2}{k}. \tag{105}$$

But in reality the number of vibrating electrons is large and they vibrate at various frequencies v_i. Equation (104) can thus be rewritten for a multi-electron system in the form of a sum over all electron frequencies

$$\alpha = \frac{e^2}{4\pi^2 m}\sum_i \frac{f_i}{v_i^2 - v^2}. \tag{106}$$

In this equation f_i represents the so-called oscillator strength, which is a measure of the degree of participation of an electron in a certain vibration and hence of the intensity of the corresponding absorption band. The sum of all f_i's equals the total number of electrons. The coefficient of static polarizability will then be expressed as follows:

$$\alpha_0 = \frac{e^2}{4\pi^2 m} \sum_i \frac{f_i}{v_i^2} = e^2 \sum_i \frac{f_i}{k_i},$$ (107)

and for the refractive index we can write

$$\frac{n^2 - 1}{n^2 + 2} = N_i \frac{e^2}{3\pi m} \sum_i \frac{f_i}{v_i^2 - v^2}.$$ (108)

This equation is termed the *dispersion equation*.

If, according to Maxwell's theory, the relationship

$$n_{\lambda \to \infty}^2 = \varepsilon,$$ (109)

holds for infinitely long waves (in a static electric field), then for permittivity ε we get an analogous equation of dispersion, viz.

$$\frac{\varepsilon - 1}{\varepsilon + 2} = N_i \frac{e^2}{3\pi m} \sum_i \frac{f_i}{v_i^2 - v^2}.$$ (110)

The natural frequencies, v_i, of materials which are non-absorbing in the visible region of the spectrum lie within the range of the shorter wavelengths (in the ultraviolet range of the spectrum) and $v_i > v$.

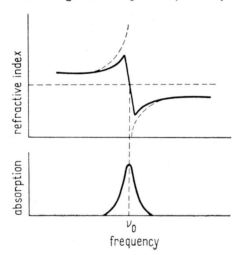

Fig. 40 Variation of the refractive index with frequency in the region of an absorption band.

With frequency v decreasing, the denominator of equation (109) will increase and the refractive index will thus grow smaller. With $v = v_0$, the refractive index would exceed all limits, which from the physical point of view would of course be impossible. If, however, account is taken of the absorption, which occurs whenever the frequencies of the electrons and those of the incident radiation resonate, the changing index of refraction can be graphically represented (after [2]), as shown in Fig. 40.

80

We can see that in the vicinity of the absorption band the values of n will vary. This phenomenon is termed *anomalous dispersion*.

So far we have considered classical theory only, which makes use of a simple model of the electron, viz. the harmonic oscillator. But, as we have already established (see Chapter 3), light is radiated, or absorbed, when the electrons pass from one energy level to another. The frequencies are thus determined by the difference between the values of energy of the two levels (i, k) which the electron passes through.

$$v_{ik} = \frac{1}{k}(E_i - E_k). \tag{111}$$

With help of quantum mechanics, it can be shown that the classical model of the harmonic oscillator can well be made use of in this case. But in place of frequencies v_i and forces f_i with one index i, we write v_{ik} and f_{ik}, thus characterizing the two energy states between which the transition takes place.

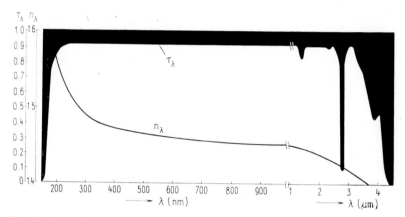

Fig. 41 Dispersion curve of quartz glass compared with its absorption spectrum.

The dispersion sum will thus contain terms representing transitions into the excited states 1, 2, 3, ...; and we can write

$$\alpha = \frac{e^2}{4\pi^2 m} \sum_i \frac{f_{io}}{v_{io}^2 - v^2}, \tag{112}$$

where the force of the oscillator, f_{io}, is brought into connection with the probability of the quantum transition

$$E_0 \to E_i. \tag{113}$$

Index $(_0)$ refers to the electronically non-excited state of the system.

Not only electrons, but also atomic nuclei, are displaced under the effect of an electric field. Of course, these displacements are much smaller than the displacement of the electrons. The nuclei have their own vibrations, which fall under the infra-red range of the spectrum.

Figure 41 compares the refractive index, n_λ, of quartz glass and its spectral transmittance τ_λ.

In agreement with the equation of dispersion (108), the index of refraction will clearly depend upon the wavelength of the radiation, and this relationship must therefore not only be respected, but in many cases it can be made use of in research into optical properties. And we can see that in the absorptive region the refractive index reaches its maximum at a wavelength of approximately 200 nm, and it drops to a minimum at the approximately 4 μm wavelength.

In the same way, the refracted ray of light, like the reflected ray, is polarized too. The ray of light is in part linearly polarized, with the plane of vibration running parallel to the plane of incidence. The degree of polarization, in its turn, depends upon the angle of incidence, and the polarizing effect will be more pronounced if the number of refractions is larger, viz. if the ray of light is refracted at more than one interface [3]. We shall deal with these questions in more detail in Chapter 4.5.

Definition of the refraction of radiation

Refraction is the term applied to a change in the direction of the radiation caused by local differences between the velocities of the propagation of the radiation in a heterogeneous medium, or occurring during the passage of the radiation through the interface between two media exhibiting different indices of refraction.

The refractive index of a medium for monochromatic λ wavelength radiation is the ratio of the velocity of the electromagnetic waves in a vacuum to the phase velocity for the given medium; thus,

$$n_\lambda = \frac{c}{v_\lambda} \quad [\mathrm{m \cdot s^{-1}/m \cdot s^{-1}}], \tag{114}$$

where c is the velocity of the propagation of radiation in a vacuum,
and v_λ = the phase velocity in the medium.

The air-related index of refraction, n, is a similar quantity, which is currently applied.

The refractive index can also be expressed as the ratio of the sine of the angle of incidence, Θ_1, to the sine of the angle of refraction, Θ_2, when the ray passes through the surface separating the vacuum from the medium (see equation (94)).

4.2.3   Dependence of the refractive index of glasses upon the wavelength of monochromatic radiation

From the dispersion equation (108) it follows that the refractive index of glasses depends upon the wavelength (the frequency) of monochromatic radiation. The dispersion curves for quartz glass and for optical glass (Figs. 42 and 43, respectively) are given as examples.

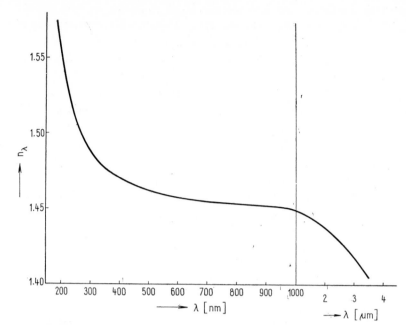

Fig. 42 Dispersion curve of quartz glass.

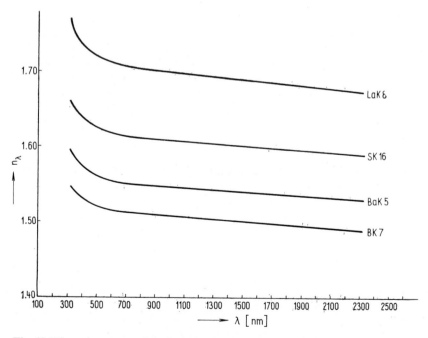

Fig. 43 Dispersion curves of the LaK 8, SK 16, BaK 5, and BK 7 optical glasses.

In agreement with the theory, we find that the index of refraction rises steeply in the region of the wavelengths where the electrons resonate with the frequency of the incoming radiation. In the ultra-violet region of the spectrum the index of refraction will thus increase owing to the resonance of the frequency of the radiation with the frequency of the electrons of $Si-O$ bonds (see Section 4.3.3).

On the other hand, in the near infra-red region of the spectrum the index of refraction decreases owing to the resonance of the frequency of the radiation with the OH groups bonded in glass; and in the region from 5 μm wavelength upwards its fall is steep owing to the resonance of the frequency of the radiation with the vibrations of the $Si-O$ bonds. With longer wavelengths, rotation will appear. These questions are dealt with in detail in Section 4.3.3.

The refractive index is currently defined for the wavelengths of $\lambda = 546.07$ nm and $\lambda = 587.56$ nm and is denoted as n_e or n_d. In Table 13 is a list of the wavelengths of radiation for which the refractive index of optical glasses is usually measured. Apart from the symbols denoting the refractive indices for the listed wavelengths, the table also quotes the sources of radiation applied.

Table 13

Refractive index	Wavelength	Discharge lamp filled with	Refractive index	Wavelength	Discharge lamp filled with
n_i	365.01	Hg	n_D	589.29	Na
n_h	404.66	Hg	$n_{C'}$	643.85	Cd
n_g	435.84	Hg	n_C	656.27	H
$n_{F'}$	479.99	Cd	n_r	706.52	He
n_F	486.13	H	n_A	769.90	K
n_e	546.07	Hg	n_S	852.11	Cs
n_d	587.56	He	n_t	1 013.98	Hg

For the wavelengths listed above, the so-called *Abbe' values*, v_e and v_d, are then computed from the relationships

$$v_e = \frac{n_e - 1}{n_{F'} - n_{C'}}, \tag{115}$$

$$v_d = \frac{n_d - 1}{n_F - n_C}. \tag{116}$$

Optical glasses are also defined by mean dispersion,

$$n_{F'} - n_{C'}, \tag{117}$$

$$n_F - n_C. \tag{118}$$

In cases which are essential for the calculations of optical systems, optical glasses are also defined by relative dispersions, which can be computed from the relationship

$$\frac{n_{\lambda_1} - n_{\lambda_2}}{n_\lambda - 1}, \tag{119}$$

where $\lambda_1 < \lambda < \lambda_2$.
Thus, for instance,

$$\frac{n_s - n_t}{n_F - n_C}; \quad \frac{n_d - n_{C'}}{n_{F'} - n_{C'}}; \quad \frac{n_e - n_d}{n_{F'} - n_{C'}}; \quad \frac{n_i - n_h}{n_{F'} - n_{C'}}. \tag{120}$$

Other partial dispersions are also made use of, e.g., $n_h - n_g$, $n_g - n_F$, $n_d - n_C$, $n_C - n_A$ etc. These characteristics define the properties of glasses applied optically.

From the dispersion curves in Figures 42 and 43 it thus follows that, with the wavelength increasing, the value of the refractive index will decrease. This effect is in full agreement with the theory, if we consider the fact that the increase in wavelength is accompanied by a decrease in the energy of that radiation (see Table 3).

The dependence of the index of refraction upon the wavelength of monochromatic radiation is taken full advantage of in optical practice, and a number of measuring methods are based upon this effect.

4.2.4   Dependence of the refractive index of glasses upon temperature

The index of refraction of glasses depends upon temperature. In dealing with these questions distinction must be made between the region of temperatures up to the transformation range and the region of temperatures within the transformation range. Within the transformation range the properties of glasses undergo changes, which are characteristic of amorphous materials. These questions are dealt with in detail in the literature listed in references [4, 6, 7, 8].

In order to explain the thermal change in the refractive index of glasses, Prod'homme [5] considers two counter-acting effects:

(a) the growth of the specific volume caused by an increase in temperature is accompanied by a drop in the refractive index;

(b) an increase in the polarizability caused by the rising temperature is accompanied by a rise in the refractive index.

The effect of these two phenomena upon the index of refraction can, according to [5], be derived from the relationship

$$n^2 = \frac{M_v + 2R}{M_v - R}, \tag{121}$$

where R stands for specific refraction,
and M_v for molar volume.

By differentiation we get

$$2n\,dn = \frac{3M_v\,dR - 3R\,dM_v}{(M_v - R)^2} = \frac{3RM_v}{(M_v - R)^2}\left(\frac{dR}{R} - \frac{dM_v}{M_v}\right). \tag{122}$$

If

$$\frac{3RM_v}{(M_v - R)^2} = \frac{(n^2 - 1)(n^2 + 2)}{3}, \tag{123}$$

then differentiating $\dfrac{dM_v}{M_v}$ and $\dfrac{dR}{R}$ with respect to temperature T we get

$$\frac{dM_v}{M_v}\frac{1}{dT} = \beta, \tag{124}$$

and

$$\frac{dR}{R}\frac{1}{dT} = \varphi = \frac{d\alpha}{\alpha}\frac{1}{dT}, \tag{125}$$

where α stands for polarizability if $R \sim \alpha$.

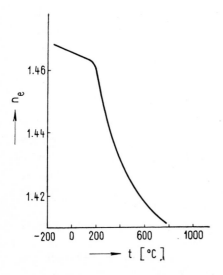

Fig. 44 Dependence of the refractive index of vitreous B_2O_3 upon temperature.

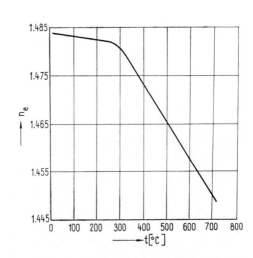

Fig. 45 Dependence of the refractive index of the $0.9\ Na_2O \cdot P_2O_5$ glass upon temperature.

Then

$$\frac{dn}{dT} = \frac{(n^2 - 1)(n^2 + 2)}{6n}(\varphi - \beta). \tag{126}$$

If the approximately constant values of the term $\dfrac{(n^2 - 1)(n^2 + 2)}{6n}$ are disregarded, then the thermal change in the refractive index will depend upon the difference

86

between the values of the thermal change in polarizability, φ, and the volume expansion, β. On the basis of an investigation covering a large range of temperatures and glasses of various types of composition, glasses can be classified into three basic groups, with the thermal change in the refractive index as criterion:

(a) $\beta > \varphi$: with the temperature increasing, the refractive index of glasses decreases. This is, for example, the case of vitreous boric oxide, some boric glasses and phosphoric glasses, organic glasses etc. Figures 44 and 45 give a graphical representation of the dependence of the refractive index upon temperature in glasses of this group. Figure 46 shows the dependence of the specific mass upon temperature in vitreous boric oxide (after [47]).

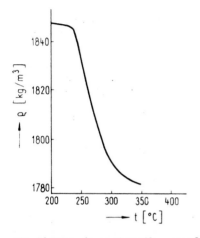

Fig. 46 Dependence of specific mass of vitreous B_2O_3 upon temperature.

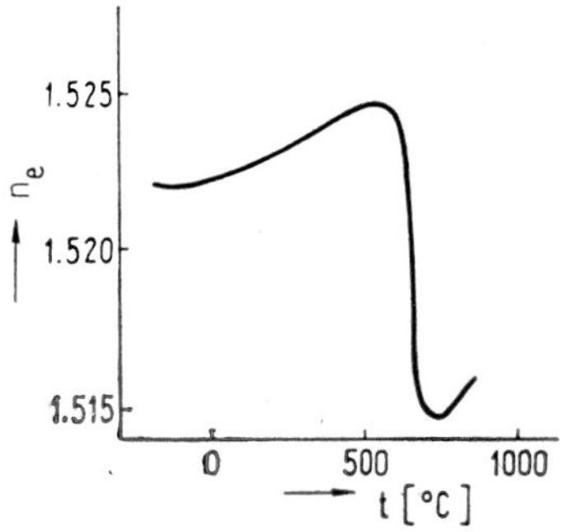

Fig. 47 Dependence of the refractive index of borosilicate glass upon temperature.

(b) $\beta \sim \varphi$: the curves of thermal dependence, β and φ, intersect, so that the coefficient $\xi = \dfrac{dn}{dT}$ is either positive or negative. The temperature-dependent change in the refractive index takes a more complex course in this case, which is characteristic, for example, of the majority of optical glasses, borosilicate glasses etc. Figures 47 and 48 show the dependence of the refractive index upon temperature of glasses in this group.

(c) $\beta < \varphi$: with increasing temperature, the refractive index of the glasses increases continually (in the transformation range, too). This applies to quartz glass, where the growth in the specific volume caused by an increase in temperature is extraordinarily small, so that the effects of polarization prevail (see Fig. 49).

In the thermal region below the transformation range the dependence of the refractive index of glasses upon temperature can be characterized, with some approximation, by a linear function of temperature. In certain glasses the refractive

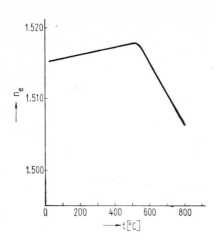

Fig. 48 Dependence of the refractive
index of BK 7 glass upon temperature.

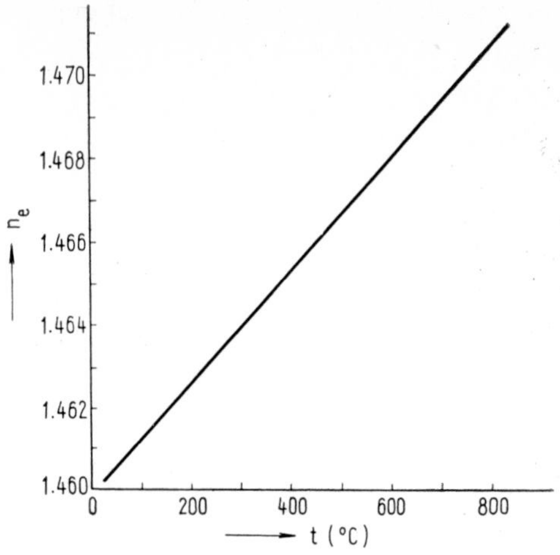

Fig. 49 Dependence of the refractive index of
quartz glass upon temperature.

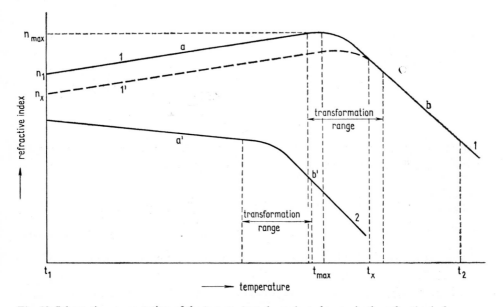

Fig. 50 Schematic representation of the temperature-dependent changes in the refractive index.

index rises in this region (Figures 47, 48, 49), while in other glasses it decreases
(Figures 44, 45). The temperature-dependent variations of the refractive index of
glasses are schematically represented in Figure 50.

Figure 51 shows the dependence of the change in the index of refraction of optical glass BK 7 in the range of temperature from $-40°$ up to $+80$ °C for three
wavelengths.

88

Let us, however, return back to Figure 50. If silicate glass (curve I in Figure 50) is cooled down from temperature t_2 at a rate which ensures that at each moment an equilibrium state is established, the index of refraction will rise up to a point, t_{max}, where the maximum value, n_{max}, is reached. In the course of further cooling the decrease in the refraction index will follow curve I, and a value n_1 will be attained at temperature t_1.

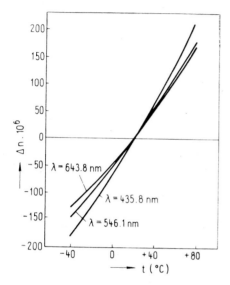

Fig. 51 Dependence of the change in the index of refraction of BK 7 optical glass upon temperature.

If, however, cooling proceeds so rapidly that the structural changes cannot follow the structural arrangement at higher temperatures, the change in the refractive index will take the course of curve I', and at temperature t_1 the refractive index of the glass will equal n_x. The index of refraction of this glass will thus be lower than in the case of a slower process of cooling. Glass of this type is termed *non-stabilized glass.*

If this non-stabilized glass is again heated to a temperature corresponding to temperatures in the transformation range its refractive index will increase until it reaches the value given by curve I in Figure 50. The duration of the thermal treatment is thus also of importance. Figure 52 shows measurements of the refractive index of non-stabilized glass as a function of the duration of the thermal treatment at constant temperature.

From the course of the curve in Figure 52 it follows that with prolonged thermal treatment, the refractive index of non-stabilized glass will rise to a maximum, viz. to an equilibrium value.

These questions are dealt with in more detail by Prod'homme [5], McMaster [7], Collyer [8], and Daniushevski [9].

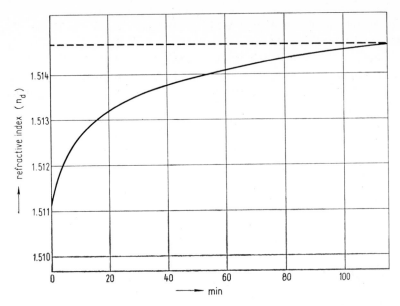

Fig. 52 Dependence of refractive index upon time at constant temperature.

4.2.5   Stabilization of the refractive index of glasses

As already mentioned above, the refractive index of glasses in the region of the transformation range depends upon the rate of cooling. The structure of glass thus depends upon the thermal history and can be characterized by the so-called *fictive temperature*, τ.

Scientifically based cooling schedules, particularly of optical glasses, were first suggested by Adams and Williamson [10], with respect to stress arising in these glasses. These authors did not, however, take account of the fictive temperature of glass. Their procedures are characterized by both a relatively long holding time in the upper temperature region and a gradually increasing rate of cooling from the upper cooling temperature to the lower cooling temperature. The transition from the lower cooling temperature to ambient temperature proceeds at a constant rate.

The first to deal with the so-called *stabilization cooling* of optical glasses was Brandt [11]. He proceeded from the fact that when glass is heated to a certain constant temperature in the cooling interval, the refractive index, n, will change continuously and will approach an equilibrium value. The period of the duration required to reach the equilibrium value depends upon the temperature.

The attainment of a uniform index of refraction for the whole volume of the glass (equal fictive temperature in the different parts of the glass) depends upon the glass being transferred from the upper part of the transformation range to the lower range in such a way that all the parts of the volume of the glass shall

90

be submitted to equal thermal treatment. In the inner parts of the volume of the glass cooling will proceed more slowly than in the surface parts; it is however essential to ensure that these differences are kept constant for the whole process of annealing.

Annealing of optical glass at a constant rate, which has been described by Daniushevski [9], Lillie and Ritland [12], is termed *rate annealing*. In this type of annealing the final index of refraction is governed by the constant rate of cooling through the transformation range. The higher the rate of cooling, the lower the

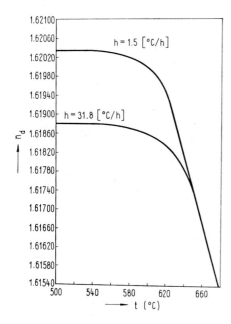

Fig. 53 Dependence of the refractive index of SK 16 optical glass upon temperature at two different rates of cooling.

refractive index attained. Figure 53 gives data on the dependence of the refractive index of SK 16 glass upon temperature for two annealing rates in the transformation range.

The dependence of the refractive index upon the logarithm of the rate of annealing is linear, the gradient of line is negative and depends upon upon the chemical composition of the glass. The linear dependence of the refractive index upon the logarithm of the rate of annealing can be taken advantage of in the modification of the refractive indices of glass. If, for a certain type of glass a refractive index, n, is required, the rate of annealing can be computed from the relationship

$$n = n_0 + \overline{m} \log \frac{h_0}{h},$$
(127)

where $\overline{m}$ is the direction,

n is the required refractive index of the glass,

n_0 is the refractive index of the glass annealed at a rate h_0 (the expected range being from 1 to 2.5 °C . h^{-1}),

and   **h** is the computed annealing rate.

The determination of the annealing rate, **h**, in the transformation range is impeded by the fact that stress may arise in the glass: so the possibility of adjusting and modifying the refractive index of glass is thus limited by the value of the direction $\overline{m}$ and by the dimensions (volume) of the glass. Permanent stress in glass is proportional to the squared thickness of the glass. These questions are dealt with more comprehensively in [13, 49].

4.2.6   Dependence of the refractive index upon the composition of the glass

The phase velocity of light passing through glass, and the refractive index too, depend upon the interaction of the electric field of that ray and the electron shells of the ions, the molecules, or the more complex structures which make up the matter of the medium, viz. glass. The higher the polarizability of the outer electrons, the lower the phase velocity of the ray of light.

In explaining the effect of the composition of glass upon the refractive index, we shall base our considerations upon theoretically proven relationships for *specific refraction*, **R**, and *molar refraction*, **R$_M$**, derived by Lorenz and Lorentz:

$$\mathbf{R} = \frac{n^2 - 1}{n^2 + 2} \frac{1}{\varrho}, \tag{128}$$

$$\mathbf{R_M} = \frac{n^2 - 1}{n^2 + 2} \frac{M}{\varrho}, \tag{129}$$

where ϱ is the specific mass (density),

and M the molecular weight.

The importance of the specific and the molar refraction thus lies in the relationships of those refractions to the inner structure of the glass.

The ratio $\dfrac{M}{\varrho}$ gives the molar volume.

Molar refraction, **R$_M$**, is proportional to polarizability α according to the relationship

$$\mathbf{R_M} = \frac{4\pi N_L}{3} \alpha, \tag{130}$$

where N_L is the Loschmidt number.

Since the velocity of the passage of the ray of light through glass depends upon the mutual interaction of the electric field of the ray and the outer electron shells of the atoms, the ions, or the molecules, the latter are subjected to induced polarization by the electric field of the ray. This polarization consists in the displacement of the centroid of the negative charge of the electron shell with respect to the positive charge of the atomic nuclei. The atom, the ion, or the molecule, thus gain an *induced dipole moment*, *p*, the magnitude of which depends upon the forces which bond the electrons to the nuclei. For the induced dipole moment, *p*, we can write,

$$p = \Sigma \frac{e^2}{k} \mathbf{E} = lq, \tag{131}$$

where $\Sigma \dfrac{e^2}{k} = \alpha$ is the deformability or polarizability,

$\quad$ e $\ $ is $\quad$ the charge of the electron,
$\quad$ k $\ $ is $\quad$ the force acting at unit distance,
$\quad$ $\mathbf{E}$ $\ $ is $\quad$ the electric field of the ray of light,
$\quad$ l $\ $ is $\quad$ the distance between the two centroids,
and $\quad q$ $\ $ is $\quad$ the effective charge in the centroids.

Table 14

Component	mol. %						
	1	2	3	4	5	6	7
BeF_2	65	60	55	50	45	45	45
AlF_3	5	10	15	20	21	21	21
KF	8	8	8	8			
MgF_2	10	10	10	10	17	17	17
CaF_2	8	8	8	8	7	7	7
SrF_2	2	2	2	2	5	5	5
LaF_3	2	2	2	2	3	3	3
CeF_4					2	2	2
H_3BO_3						2	4
n_D	1.3300	1.3348	1.3380	1.3400	1.3565	1.3578	1.3593
v_D	108.0	105.1	100.3	97.6	99.8	99.3	99.6

Molar refraction is thus a measure of deformation (polarizability) and its value is approximately the sum of the ionic refractions, R_i. Owing to their looser structure, the anions which are present in the glass are more polarizable than the cations. The anions can be arranged in a sequence ordered according to the degree of their polarizability, viz.

$$F^- < OH^- < Cl^- < O^{2-} < S^{2-} < Se^{2-} < Te^{2-}. \tag{132}$$

Thus anions F^- are the least polarizable, and anions Te^{2-} the most polarizable. The lower polarizability of F^- as compared with O^{2-} is also manifested in the reduced molar refraction and a lower index of refraction of, for example, fluoride or fluorophosphate glasses as against oxide glasses.

Table 14 lists the results of the measurement of the refractive index and the Abbe values for pure fluoride glasses as compared with glasses with added boric acid [50].

Chalcogenide glasses, on the other hand, containing such anions as S^{2-}, Se^{2-}, and Te^{2-}, have a substantially higher refractive index than oxide glasses as shown in Figures 54 and 55 (after [45, 46]).

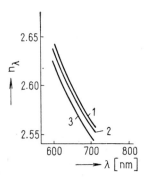

Fig. 54 Dependence of the refractive index of As_2S_3 upon wavelength: *1* — rate of cooling 15 K/hour; *2* — rate of cooling 25 K/hour; *3* — rate of cooling 20 000 K//hour.

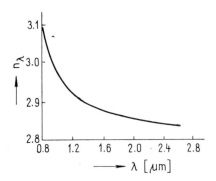

Fig. 55 Dependence of the refractive index of 95 As_2Se_3 . 5 As_2Te_3 upon the wavelength.

Owing to the high polarizability of S^{2-}, Se^{2-}, and Te^{2-}, the chalcogenide glasses are not transmissive in the visible region of the spectrum, but they are fully transmissive in infra-red.

The O^{2-} anion, which is contained in the majority of commercially manufactured glasses, lies in the middle of the above ordered sequence.

Considering the oxygen anion, we find that its ionic refraction is affected by the polarizing influence of the modifying cations, viz. by the intensity of cationic field z/a^2 (where z is the charge of the cation and a the distance between the ions).

As the number of oxygens increases, the ionic refraction will increase. Figure 56 gives data on the dependence of molar refraction R_M and molar volume M_v upon the number of oxygen anions in MO_n, according to [14].

The number of anions surrounding the cation defines the symmetry of the latter. In oxide glasses, the general supposition is tetrahedral symmetry — coordination number 4, and octahedral symmetry — coordination number 6. The dimensions and the coordination number of the cations thus influence polarizability and molar refraction.

The molar refraction of glass can be calculated from ionic refractions, the values of which for some ions are listed, together with the radii, the likely coordination numbers, and the intensities of the field z/a^2, in Table 15.

In Table 16 the reader will find data on the ionic refractions of O^{2-} in certain silicates, according to [15].

In quartz glass, in which partially covalent and partially ionic bonds occur between $Si-O$ (bridging oxygens), polarizability is rather low, owing to the rigidity of the bonds, and quartz glass thus exhibits a relatively low index of refraction.

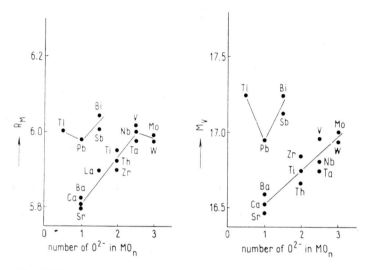

Fig. 56 Dependence of molar refraction, R_M, and molar volume, M_v, upon the number of oxygen anions in MO_n.

The introduction of, for example, Na^+ ions will destroy the bonds of the oxygen bridges, non-bridging oxygens will appear bonded to one atom of silicon only, and the bonds between oxygen and sodium will mostly be of ionic character (much weaker than those between oxygen and silicon). These non-bridging oxygens manifest higher polarizability, which increases with the concentration of Na^+. Binary glasses containing potassium behave likewise. Here, too, the introduction of K^+ ions will destroy the bridging oxygens. The introduction of Li^+ ions has a different effect from that of the Na^+ and K^+ ions; the difference must be attributed to the small ionic radius and hence to the higher force fields.

The increase in polarizability is accompanied (according to equation (130)) by an increase in molar refraction and hence also by an increase in the refractive index. Figure 57 shows the dependence of n_D upon the concentration of sodium, potassium, and lithium oxides (mol. %) in R_2O-SiO_2 binary glasses.

The refractive index depends not only upon polarizability but also upon molar volume $\dfrac{M}{\varrho}$.

Table 15

Ion	Radius of ion coordination number [6] [nm]	Probable coordination number related to O^{2-} ion	Ionic refraction R_i	Field intensity $[z/a^2]$
Li^+	0.068	4, 6	0.08	0.23
Na^+	0.098	6, 8	0.47	0.19
K^+	0.133	6, 10, 12	2.24	0.13
Rb^+	0.149	10, 12	3.75	0.12
Cs^+	0.165	12	6.42	0.10
Ti^+	0.149	—	10.0	0.14
Be^{2+}	0.034	4	0.03	0.86
Mg^{2+}	0.074	4, 6	0.26	0.51
Ca^{2+}	0.104	6, 8	1.39	0.35
Sr^{2+}	0.120	6, 8, 12	2.56	0.27
Ba^{2+}	0.138	8, 12	4.67	0.24
Zn^{2+}	0.083	4, 6	0.7	0.59
Cd^{2+}	0.099	6, 8	2.8	0.44
Pb^{2+}	0.126	6,8	7.1	0.34
Mn^{2+}	0.091	4, 6, 8	1.7	0.48
Fe^{2+}	0.080	4, 6, 8	1.2	0.52
Co^{2+}	0.078	4, 6	1.1	0.53
Ni^{2+}	0.074	4, 6	1.0	0.55
Cu^{2+}	0.080	4, 6	1.1	0.53
B^{3+}	0.020	3, 4	0.006	1.62 1.45
Al^{3+}	0.057	4, 6	0.14	0.97
Ga^{3+}	0.062	4, 6	0.6	1.08
In^{3+}	0.092	6, 8	2.0	—
Sc^{3+}	0.083	6	0.9	0.65
Y^{3+}	0.097	6, 8	1.7	—
La^{3+}	0.104	8	3.5	0.43
Fe^{3+}	0.067	4, 6	1.1	0.91
As^{3+}	0.069	4, 6	1.7	1.00
Sb^{3+}	0.090	4, 6	2.8	0.73
Bi^{3+}	0.120	6, 8	3,8	0.62
Si^{4+}	0.039	4	0.08	1.56
Ge^{4+}	0.044	4, 6	0.4	1.75
Sn^{4+}	0.067	4, 6	1.5	1.13
Ti^{4+}	0.064	4, 6	0.6	1.25
Zr^{4+}	0.082	6, 8	1.1	0.84
Hf^{4+}	0.082	6, 8	1.1	—
Th^{4+}	0.095	6, 8	3.9	.0.68
P^{5+}	0.035	4	0.05	2.08
V^{5+}	0.040	4	0.4	1.85
Nb^{5+}	0.066	4, 6	0.9	—
Ta^{5+}	0.066	4, 6	0.8	—

Table 15 (cont.)

Ion	Radius of ion coordination number [6] [nm]	Probable coordination number related to O^{2-} ion	Ionic refraction R_i	Field intenisty $[z/a^2]$
Cr^{6+}	0.035	4	0.3	2.40
Mo^{6+}	0.065	4, 6	0.7	2.15
W^{6+}	0.065	4, 6	0.6	—
F^-	0,133	—	2.44	—
O^{2-}	0.136	—	6.95	—
S^{2-}	0.182	—	22.7	—
Se^{2-}	0.193	—	28.8	—
OH^-	—	—	4.85	—

Table 16

Silicate	R_i [cm³]	Silicate	R_i [cm³]
Be_2SiO_4	3.35	Sr_2SiO_4	4.67
SiO_2 (quartz)	3.55	Ba_2SiO_4	4.97
Mg_2SiO_4	3.83	$BaSiO_3$	4.55
$LiAlSiO_4$	4.01	$Ba_2Si_3O_8$	4.3
Ca_2SiO_4	4.53	$BaSi_2O_5$	4.1
$NaAlSiO_4$	4.03		

From the equation

$$n^2 = \frac{1 + 2Y}{1 - 2Y}; \qquad Y = \frac{4\pi N_L}{3} \alpha \frac{\varrho}{M}, \tag{133}$$

it follows that the refractive index increases when the molar volume decreases, viz. when the structure gets "denser". In the diagram of Figure 58 are plotted the molar volumes of binary glasses of compositions identical to those quoted for the refractive index in Figure 57.

Figure 59 shows the effect of composition on the refractive index of binary glasses of the basic composition, viz. 80 wt. % SiO_2 and 20 wt. % NaO_2; aluminium oxide, boric oxide, magnesium oxide, zinc oxide, lead oxide, and calcium oxide have been introduced at the expense of SiO_2.

In most cases, the increase in the refractive index is accompanied by an increase in specific mass (density), as shown in Table 17.

There are, however, cases, e.g., in the $Na_2O-P_2O_5$ system, where the increase in specific mass is accompanied by a decrease in the index of refraction.

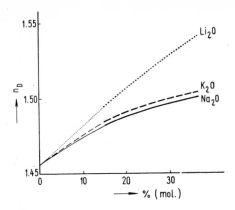

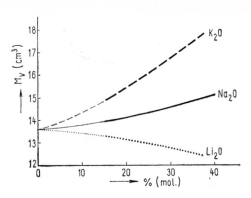

Fig. 57 Dependence of the refractive index upon the concentration of Na₂O, K₂O, and Li₂O in R₂O-SiO₂ binary glass.

Fig. 58 Dependence of molar volume M_v upon the concentration of Na₂O, K₂O, and Li₂O in R₂O-SiO₂ binary glass.

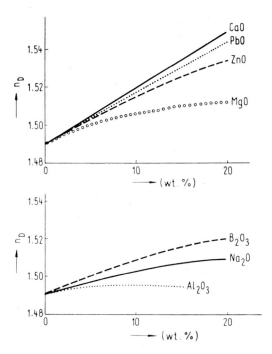

Fig. 59 Dependence of the refractive index upon the concentration of Al₂O₃, B₂O₃, MgO, ZnO, PbO, CaO, and Na₂O in a binary glass of 80 wt. % SiO₂ and 20 wt. % Na₂O (with components introduced at the expense of SiO₂).

Süsser and Fanderlik [30] made a study of the effect of tungsten trioxide and strontium, cadmium, zirconium, thorium, and lanthanum oxides upon the refractive index of a glass of the following composition: 31.2 wt. % SiO₂, 17.9 wt.% B₂O₃, 48.6 wt. % BaO, 1.4 wt. % Al₂O₃, 0.4 wt. % Sb₂O₃, 0.5 wt. % As₂O₃. The quoted oxides were introduced at the expense of SiO₂. The results of the measurements are presented in Figures 60, 61, 62, 63, 64, 65.

Table 17

Weight %		n_D	$n_F - n_C$	$\varrho\ [kg/m^3]$
Na₂O	B₂O₃			
1.82	98.18	1.4715	0.0079	1890.1
3.66	96.34	1.4772	0.0079	1933.2
5.66	94.34	1.4828	0.0080	1974.4
7.38	92.62	1.4876	0.0081	2011.2
9.33	90.67	1.4915	0.0081	2046.6
10.90	89.10	1.4940	0.0081	2075.2
12.78	87.22	1.4963	0.0082	2105.3
16.53	83.47	1.5000	0.0082	2156.7
22.23	77.77	1.5072	0.0083	2249.3
25.91	74.09	1.5140	0.0083	2314.1
31.63	68.37	1.5174	0.0084	2375.5

Glasses showing a particularly high index of refraction have been the object of studies made by Faulstich [36], Geffcken [35], Izumitani [14], Jahn [37], Meinecke [39], Armistead [51] and others. Figure 66 (after [36]) shows the glassforming region in a three-component $B_2O_3 - La_2O_3 - Ta_2O_5$ system as well as values of the dependence of the refractive index, n_d, upon Abbe's value v_d; Figure 67 is a similar diagram for the $17\ B_2O_3 - 15\ Nb_2O_5 - Ta_2O_5 - La_2O_3 - ThO_2$ system.

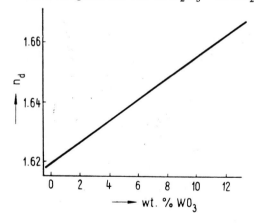

Fig. 60 Dependence of the refractive index upon the concentration of WO₃ in a glass of the following composition: 31.2 wt. % SiO₂, 17.9 wt. % B₂O₃, 48.6 wt. % BaO, 1.4 wt. % Al₂O₃, 0.4 wt. % Sb₂O₃, 0.5 wt. % As₂O₃ (with the exchange at the expense of SiO₂)

According to Izumitani [14], the dispersion of glasses depends upon the absorptive properties of the glasses in the ultra-violet region of the spectrum, which is in agreement with the equation of dispersion (108). Higher values of dispersion are thus obtained for the glasses in which the absorption edge is displaced into the visible region of the spectrum, and vice versa. Figures 68 and 69 show dispersion in the SF 7 and SF 59 optical glasses.

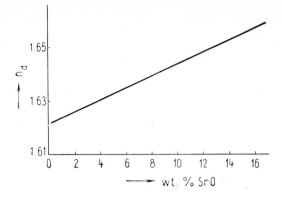

Fig. 61 Dependence of the refractive index upon the concentration of SrO in a glass of the composition quoted in the legend to Figure 60.

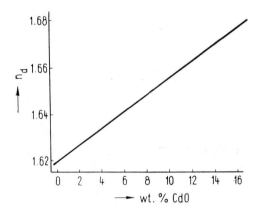

Fig. 62 Dependence of the refractive index upon the concentration of CdO in a glass of the composition quoted in Figure 60.

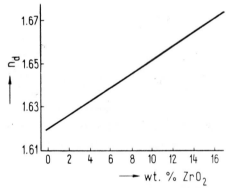

Fig. 63 Dependence of the refractive index upon the concentration of ZrO_2 in the glass of the composition quoted in Figure 60.

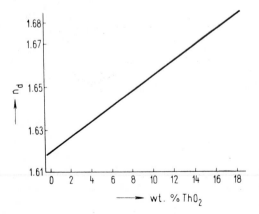

Fig. 64 Dependence of the refractive index upon the concentration of ThO_2 in the glass of the composition quoted in Figure 60.

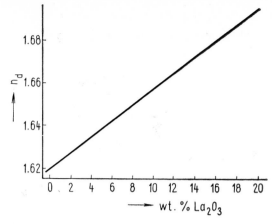

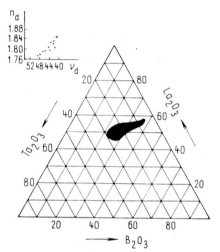

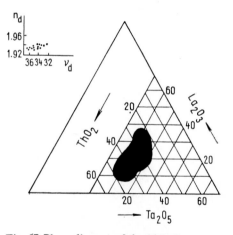

Fig. 65 Dependence of the refractive index upon the concentration of La_2O_3 in a glass of the composition quoted in Figure 60.

Fig. 66 Phase diagram of the B_2O_3 — La_2O_3 — Ta_2O_5 system and representation of attainable refractive ndices and Abbe value (dark area).

Fig. 67 Phase diagram of the $17\ B_2O_3$ — $15\ Nb_2O_5$ — Ta_2O_5 — La_2O_3 — ThO_2 and representation of attainable refractive indices and Abbe value (dark area).

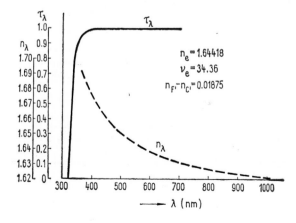

$n_e = 1.64418$
$\nu_e = 34.36$
$n_F - n_C = 0.01875$

Fig. 68 Dependence of refractive index and transmittance of SF 7 optical glass upon wavelength.

In Figure 70 (after Fanderlik [52]) we see the dependence of the refractive index upon changes in the concentration of the components introduced into the parent glass 13.5 wt. % K$_2$O, 26 wt. % PbO, 60.5 wt. % SiO$_2$, sodium oxide having been

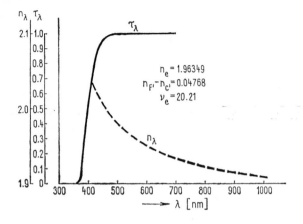

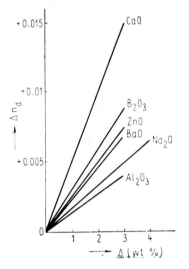

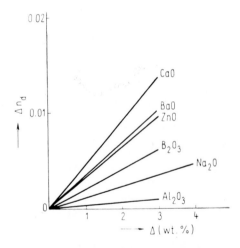

Fig. 69 Dependence of refractive index and transmittance of SF 59 optical glass upon wavelength.

Fig. 70 Dependence of the change in the refractive index upon the change in the concentration of components introduced into the K$_2$O—PbO—SiO$_2$ glass (with 26 wt. % PbO).

Fig. 71 Dependence of the change in the refractive index upon the change in the concentration of components introduced into the K$_2$O—PbO—SiO$_2$ glass (with 32 wt. % PbO).

introduced at the expense of potassium oxide, the other components at the expense of silicon oxide.

Similar results for the parent glass of 12 wt. % K$_2$O, 32 wt. % PbO, and 56 wt. % SiO$_2$ are shown in Figure 71.

The dependence of the refractive index and the dispersion of glasses upon the composition has interested a number of scientists and researchers, particularly as far as the properties of optical and special optical glasses are concerned. Should the reader be interested in this broad subject, we would refer him to the references at the end of this chapter.

4.2.7   Methods of measurement

The refractive index of glasses is measured by refractometers. These instruments are often based on the principle of total internal reflection at the critical angle. The refractive index is usually measured by *Abbe's refractometer* or *Pulrich's refractometer* and Hilger – Chance   V-refractometer. In more accurate measurements, use is made of *spectrometers*, also called *goniometers*, e.g. the Moore Master 1440 Precision Index, the Hilger and Watts TA5 photo-electric auto-collimator with strip recorder, the Guild – Watts spectrometer, the Moore 1440 small angle divider, etc.

For the measurement of the differences in the refractive indices various methods are applied, the most important of which are *Obreim's* and *Zakharevski's methods*, the method of measurement by *Young's interference refractometer,* etc. The refractive index of small-size specimens of glasses can be determined by the *Becke line method* or by a number of other methods.

Measuring the refractive index using Abbe's refractometer

Abbe's refractometer consist of a glass prism, K, the refractive index of which, n_0, ranges from 1.6 to 1.8; the prism can thus be exchanged. On the upper surface of the prism, the edges of which have been slightly bevelled, is placed a small specimen plate, G, of the glass to be measured (n) (see Fig. 72).

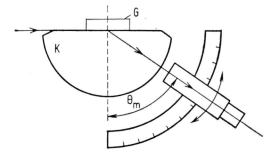

Fig. 72 Measurement of refractive index using Abbe's refractometer.

The surface of the glass specimen adjoining prism K must be polished, and the contact between the two adjoining surfaces made more perfect with the help of an immersion liquid. The glass specimen is illuminated from one side by monochromatic radiation. The ray of light falls onto the boundary between the two indices of refraction, n, and n_0, from different directions within the limits of the largest possible angle of incidence (right angle). The light is refracted into prism K at the critical angle. The latter (Θ_m) is measured by a telescope and read from the graduated circle of the instrument. The measurement is thus based on the dependence of critical angle Θ_m upon the relative indices of refraction of the two media.

If the absolute refractive index of the prism, $\bar{n}_0$, is known, the absolute refractive index of the measured glass, $\bar{n}$, can be computed from the relationship

$$\bar{n} = \bar{n}_0 \sin \Theta_m. \tag{134}$$

Refractive index $\bar{n}_0$ can be determined from the measurement of the critical angle for air, Θ_{m0},

$$n = \bar{n}_0 \sin \Theta_{m0}, \tag{135}$$

hence

$$\bar{n} = n \frac{\sin \Theta_m}{\sin \Theta_{m0}}. \tag{136}$$

If, however, the absolute refractive index of the prism, $\bar{n}_0$, is known for the monochromatic light applied, there is no need to make that measurement.

The accuracy of the measurement of the refractive index by this method is within the limits of $\pm 2 \cdot 10^{-3}$. For a more detailed description of the method see reference [25].

Measuring the refractive index using Pulfrich's refractometer

Pulfrich's refractometer consists of a set of interchangeable measuring prisms, a sighting telescope, a micrometer, and two condensers. The prisms used have different indices of refraction, the range of which sets the limits within which the measurements can be made. Table 18 lists the different prisms usually used, together with their respective ranges of measurement of the refractive indices.

Table 18

Refractive index of prism	Range of measurement of refractive index
1.61	1.61 and lower
1.73	1.56 to 1.72
1.90	1.70 and higher

A schematic diagram of Pulfrich's refractometer can be seen in Figure 73. The larger side of the specimen of glass under measurement (the dimensions being approximately $20 \times 20 \times 10$ mm) must be optically polished with an accuracy of two (interference) fringes, and one of the lateral sides with an accuracy of five fringes, to one centimeter. The right angle enclosed by these two sides must be kept within an accuracy of $\pm 10'$. A drop of immersion liquid is applied to the larger, optically polished side of the specimen; the refractive index of the immersion liquid must, of course, be higher than the refractive index of the glass under examination. Light falling parallel to the interface between the prism of the refractometer and the specimen to be measured is refracted into the prism at the critical angle Θ_m. At the interface between the prism and air the angle of refraction is Θ; it can be read from the scale of the refractometer.

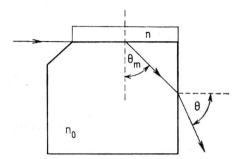

Fig. 73 Measurement of the refractive index using Pulfrich's refractometer.

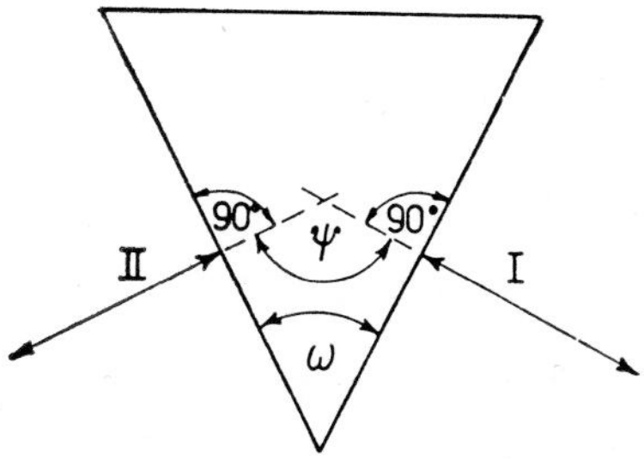

Fig. 74 Measurement of the refracting angle of a prism.

If the refractive index of the prism of the refractometer is denoted by n_0, and angle Θ is read from the scale with the cross hairs set on the boundary between light and shade, then n_λ denoting the refractive index of the glass under examination can be computed from the following equation:

$$n_\lambda = \sqrt{n_0^2 - \sin^2 \Theta}. \tag{137}$$

This air-related index of refraction is referred to as the relative index of refraction.

The method of measuring the refractive index described above [25] yields results within an accuracy of $\pm 5 \cdot 10^{-5}$.

Measuring the refractive index by spectrometer-goniometer

The application of this method requires accurate knowledge of the angle of refraction of the prism, ω, of the glass to be measured. Owing to the fact that the telescope of the spectrometer is equipped with a Gauss eyepiece, the angle of the prism is most suitably ascertained with the help of a perpendicular mirror reflection of the cross hairs. The swivel table upon which the prism of the glass under measurement is placed, is locked in position and the telescope is sighted in direction I (see Figure 74), normal to one of the faces of the prism, so that the image of the vertical hair mirrored on the refracting face shall coincide with its image in the eyepiece.

Position ψ_1 is read from the vernier of the graduated circle. The telescope is then moved to position *II*, and again the position of coincidence of the two images is found and read (ψ_2). The angle of refraction

$$\omega = 180° - \psi, \tag{138}$$

where

$$\psi = \psi_2 - \psi_1. \tag{139}$$

Thus, if the refractive angle, ω, is known, the respective refractive index for the selected length of radiation can be computed from equation

$$n_\lambda = \frac{\sin\frac{1}{2}(\omega + \delta_m)}{\sin\frac{1}{2}\omega}, \tag{140}$$

where δ_m is the minimum deviation of the ray emerging from the prism from its path of incidence (see Figure 75). The value of δ_m can be calculated from equation

$$\delta_m = \frac{1}{2}(\gamma_2 - \gamma_1), \tag{141}$$

where γ_1 denotes position (*I*) of reading on the graduated circle,
and γ_2 denotes position (*II*) of reading on the graduated circle with the prism rotated by 180°.

The accuracy of the measurement of the refractive index by this method is $\pm 5 \cdot 10^{-6}$, provided the angles are read within an accuracy of ± 0.5 angular second. The range of the values of the refractive index measurable in this way is virtually unlimited, the only prerequisite is to see that the active faces of the prism are properly cut and optically polished.

Hilger–Chance's refractometer

The refractive index is measured by an angle at which a ray of light deviates when passing through a prism block, the V-block and the specimen. The V-block consists of two glass prisms—one being a complete 45° prism and the other a 45° prism with one end accurately truncated. This prisms are joined by heat treatment to form a single block with a V-shaped niche at the top, the sides of this niche being at right angle.The specimen is roughly prepared as a right-angle prism and then placed in the niche. A contact fluid between the specimen and the V-block is used to compensate for surface irregularities.

Measuring the difference between the indices of refraction by Obreim's method

The measurement is based upon the effect of the contours of the glass fragments placed in an immersion liquid becoming invisible on being exposed to monochromatic radiation of the wavelength for which the refractive indices of the liquid and the fragmented glass coincide. Thus, if fragments of the glass to be measured and a reference plate the refractive index of which has been determined before, are immersed in the proper type of liquid, changing the wavelength of the monochromatic radiation continuously will cause the contours of one of the two pieces of glass to disappear.

The difference between the refractive index of the glass specimen and the refractive index of the reference plate must, however, not be very large. Alongside the functional edge of the reference plate (see Fig. 76) an interference pattern will be observable made up of light and dark parallel fringes.

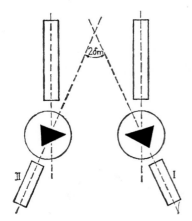

Fig. 75 Measurement of the index of refraction of glass by prism spectrometer.

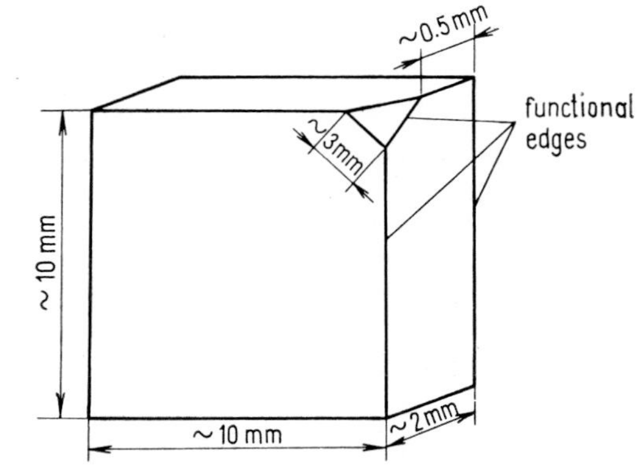

Fig. 76 Schematic diagram of the measurement of the difference between the indices of refraction by Obreim's method.

For a certain wavelength, λ_0, the refractive index of the immersion liquid and the refractive index of the reference plate will coincide, the interference pattern will disappear, and the contours of the plate will become visible. The contours of the glass specimen to be measured, the refractive index of which is different from the refractive index of the reference plate, will, however, not disappear. If disappearance of the contours of the specimen glass is to be achieved, the wavelength of the monochromatic radiation, λ, will have to be changed; and if this change is continuous from λ_0 to λ, the functional edge of the reference plate will be crossed by interference fringes. The difference between the refractive index of the reference plate and the refractive index of the specimen of the glass under examination, Δn, is given by the number of fringes, k, and determined by the relationship

$$\Delta n = n_0 - n = \frac{k\lambda}{l},$$ (142)

where Δn is the difference between the refractive index of the reference plate and the index of refraction of the specimen of glass,

n_0 is the refractive index of the reference plate,

n is the refractive index of the glass specimen,

k is the number of fringes that cross the edge of the reference plate as the wavelength changes from λ_0 to λ,

λ is the wavelength of monochromatic light at which the contours of the measured specimen of glass disappear [nm],

and l is the thickness of the reference plate [nm].

The refractive index of the glass specimen, n, can be computed from the relationship

$$n = n_0 - \Delta n, \tag{143}$$

with due consideration of the sign.

This method, which is described in more detail in [53], can be applied in cases where the difference between the indices of refraction, Δn, does not exceed $\pm 5 \cdot 10^{-3}$. The accuracy of the method is $\pm 1 \cdot 10^{-4}$.

Measuring the difference between the refractive indices by Zakharevski's method

Zakharevski's method is virtually a comparison between two glasses the difference in the refractive indices of which does not exceed $1 \cdot 10^{-3}$. Attention is paid to the deformations of the diffraction pattern evoked by the optical system of an interference refractometer. The magnitude of the path difference is given by

$$\Delta = l \, \Delta n, \tag{144}$$

where Δ stands for the path difference,

l is the thickness of the specimen of the glass to be measured,

Δn is the difference between the indices of refraction of the glasses compared.

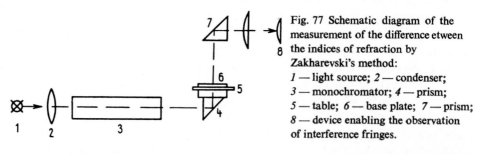

Fig. 77 Schematic diagram of the measurement of the difference etween the indices of refraction by Zakharevski's method: *1* — light source; *2* — condenser; *3* — monochromator; *4* — prism; *5* — table; *6* — base plate; *7* — prism; *8* — device enabling the observation of interference fringes.

Owing to the fact that the difference between the refractive indices, Δn, depends upon the wavelength of monochromatic radiation, a continuous change in the wavelength will be accompanied by continuous deformation of the diffraction pattern. The measuring procedure is schematically diagrammed in Figure 77.

If the path difference, $l \, \Delta n$, is equal to a whole number of wavelengths of the radiation applied and if one of those wavelengths is λ_1, a change in this wavelength will cause a shift of the diffraction maximum. If the absolute value of Δn increases, or decreases, the diffraction maximum (the lightest fringe) will move to that part of the viewing field in which the specimen exhibits a higher, or lower, index of refraction.

If λ_2 is the wavelength at which this maximum reaches the interface, the pasages of the two adjoining maxima through the interface are characterized by equations

$$l\, \Delta n_1 = h'\lambda_1, \tag{145}$$

and

$$l\, \Delta n_2 = (h' + l)\, \lambda_2, \tag{146}$$

where h' is an integer,
and for λ_1 $\Delta n_1 = \Delta n$.
For wavelength λ in the interval $\lambda_2 - \lambda_1$
we have

$$l\, \Delta n = (h' + a)\, \lambda, \tag{147}$$

where $0 < a < l$ is termed the fractional order of interference.

If λ_d is the wavelength for which Δn is measured, h' and a of equation (147) must be known. Parameter a is determined by linear interpolation, viz.

$$a = \frac{\lambda_d - \lambda_1}{\lambda_2 - \lambda_1}. \tag{148}$$

h' is obtained by auxiliary measurement of the difference between the indices of refraction, Δn, with the help of Pulfrich's refractometer for wavelength λ_d the accuracy equalling $\pm 3 \cdot 10^{-5}$. With the thickness of the plate ranging between 5 and 9.5 mm, and the difference between the refractive indices not exceeding $1 \cdot 10^{-3}$, the approximate value of the order of interference computed from

$$m = \frac{l\, \Delta n}{\lambda_d} \tag{149}$$

cannot differ from the precise value by more than $\varepsilon = \pm 0.4$ fringe. The order is thus known both in the form $h' + a$ and enclosed in the interval $m \pm \varepsilon$, where $0 < \varepsilon < 0.4$. The integral part of the order, h', will thus be allotted values allowing $h' + a$ to lie within the interval $m \pm \varepsilon$. Δn is then computed from equation

$$\Delta n = \frac{(h' + a)\, \lambda_d}{l}. \tag{150}$$

The measurement is either with, or without, the auxiliary plate, which serves to ascertain extremely small differences between the refractive indices. The accuracy of the method is $\pm 1 \cdot 10^{-5}$. The application of the method is mostly in the field of homogeneity checks of the refractive indices of glasses.

Measuring the differences between indices of refraction by Young's interference refractometer

This method is applicable in cases where the difference between the indices of refraction does not exceed the value of $1 . 10^{-3}$. The zero position of the compensator of the interference refractometer must first be ascertained, viz. with the specimen of the glass under examination being excluded. The difference between the indices of refraction for λ_d of two cement-bonded specimens is then measured with the help of Pulfrich's refractometer with the accuracy of $\pm 3 . 10^{-5}$, and the thickness of these specimens with the accuracy of ± 0.03 mm.

In the interference refractometer we then equalize the possible deviation in one part of the bonded specimens and read the value for B_{01}. After compensating in the second position (the other part of the bonded specimens), we read the value for B_{02}.

If the reading taken after compensation is different, the arithmetic mean is computed, viz.

$$B_0 = \frac{1}{2}(B_{01} + B_{02}),$$ (151)

representing zero measuring position.

The specimen is then placed in such a position that both the bonded parts are subjected to measurement. After compensation, this position will furnish us with a value for B.

Then we read the values for $B - B_0$ and find the number of fringes, k_1, making use of the calibration table. Account must be taken of whether $B \gtrless B_0$.

Substituting the recorded values into

$$\Delta n = \frac{\lambda_d(k_1 + m)}{l},$$ (152)

where $m = 0, \pm 1, \pm 2, \pm 3, ...,$ we get a number of feasible results for Δn corresponding to several orders and differing by several units. The correct value is the one that lies within, or in the nearest vicinity of, the interval

$$\Delta n \pm 3 . 10^{-5}.$$ (153)

ΔB is proportional to the compensated path difference $(B - B_0)$, viz. to the number of fringes, k_1. The function $k_1 = f(\Delta B)$ for λ_d and for the two directions of the shift is quoted in the calibration table. The shift of the k_1 interference effect is determined except for a small additive integer constant $m = 0, \pm 1, \pm 2, ...;$ and according to

$$k_1 = \frac{l \Delta n}{\lambda},$$ (154)

the determination of the difference between the indices of refraction is not unique,

$$\Delta n = \frac{\lambda_d(k_1 + m)}{l}.$$ (155)

The ambiguity can be removed by the above-mentioned measuring procedure with the help of Pulfrich's refractometer, which will enclose the correct value of the difference within the interval $\Delta n \pm 3 . 10^{-5}$.

With the thickness of the specimen ranging from 5 to 8.5 mm and the upper limit of the difference between the refractive indices of the cement-bonded glasses equalling $1 . 10^{-3}$ (the fraction part of the order being determined with the accuracy of $\pm 1/30$), there exists one value of the additive constant, m, for which Δn lies within that interval. This Δn is then the desired result of the measurement. The accuracy of the measurement is $\pm 1 . 10^{-6}$, and the method is used for measuring the homogeneity of the refractive indices of glasses.

Measuring the refractive index with the help of the Becke line

For the measurement of the refractive index of small-sized specimens (of glass fragments, crushed glass etc.) use can be made of a method based upon a comparison of the refractive index of a glass with an immersion liquid. The so-called Becke line serves as an indicator. Its appearance is conditional upon the specimen in the microscope being illuminated by a beam of rays converging at an extremely acute angle. This can be achieved by reducing the aperture of the condenser or by lowering the latter a little.

Current use is made of a lens the magnifying ratio of which is 1 : 10. The rise of the Becke line can be seen in Figure 78.

If an immersion liquid is used the index of refraction of which is lower than the refractive index of the glass under examination, the light line contouring the specimen of glass will move inside the limits of the fragment when the tube of the microscope is raised (see Fig. 78a). Lowering the tube of the microscope will cause the Becke line to move outside the limits of the fragment into the immersion liquid.

If an immersion liquid is applied the refractive index of which is higher than the refractive index of the glass to be measured, the direction of the movement of the Becke line is outside the fragment of glass, into the immersion liquid when the tube is raised. (see Fig. 78b).

If use is made of a set of immersion liquids the refractive indices of which have already been determined (it is, of course, also important to know the temperature at which the refractive indices of the immersion liquids were recorded), and the temperature is kept constant, the index of refraction of the glass can be determined in such a way that two immersion liquids are selected the refractive indices of which are slightly different from the refractive index of the specimen glass under examination, one of them being higher, the other lower; mixing the two liquids in in a certain ratio can produce a mixture the refractive index of which will be identical with the refractive index of the glass to be examined, and the Becke line will not appear at all. High accuracy measurements make use of monochromatic radiation; their tolerance is $\pm 3 . 10^{-3}$.

Another application of this method is to select an immersion liquid whose refractive index is higher than the refractive index of the glass to be measured, and to raise the temperature continuously until coincidence is reached between the index of refraction of the liquid and the refractive index of the glass. It is, o course, essential to know the dependence of the refractive index of the immersion liquid upon temperature.

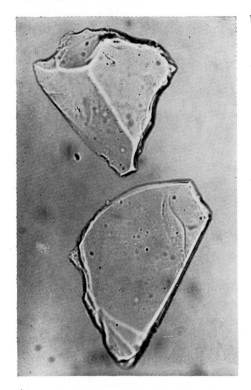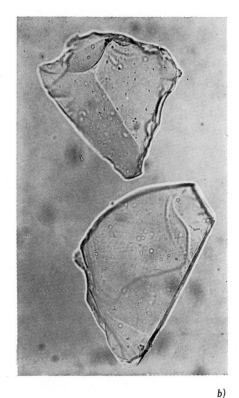

a) *b)*

Fig. 78 Movement of the Becke line in glass fragments:
(a) raising the tube moves the line inside the limits of the fragment ($n_{\text{immersion}} < n_{\text{glass}}$);
(b) lowering the tube moves the line outside the limits of the fragment ($n_{\text{immersion}} > n_{\text{glass}}$).

In Table 19 the reader will find a list of the applicable immersion liquids. Some of them are toxic, or even combustible and explosive, so that applying them involves observing the respective hygienic and safety regulations.

Measuring the refractive index of glasses at temperatures above the softening point.

The measurement is made in a special measuring cell constructed of prisms of quartz glass and placed in an electrically heated oven (see Figure 79 presenting a schematic representation of the device).

Table 19

Immersion liquid	Refractive index n_D [20 °C]	Immersion liquid	Refractive index n_D [20 °C]
methyl alcohol	1.329	benzene	1.502
water	1.333	o-oxylene	1.507
ethyl ether	1.356	trimethylene bromide	1.51
acetone	1.359	dimethyl phthalate	1.515
ethyl alcohol	1.361	cedar oil	1.516
n-propanol	1.385	chlorobenzene	1.525
amyl alcohol	1.40	ethylene bromide	1.537
isoamyl valerate	1.413	methyl salicylate	1.538
glycol	1.432	Canada balsam	1.542
chloroform	1.446	nitrobenzene	1.552
paraffin	1.448	monobromated benzol	1.560
tetrachloromethane	1.463	aniseed oil	1.56
olive oil	1.47	o-toluidine	1.573
glycerol	1.473	aniline	1.586
turpentine oil	1.475	m-chloraniline	1.593
paraffin oil	1.475	monoiodobenzol	1.621
castor oil	1.478	chinoline	1.628
tetrachloroethane	1.49	carbon disulphide	1.628
iso-propylbenzene	1.493	α-mono- bromonaphthalene	1.658
toluene	1.495	methylene iodide	1.742
xylene (mixture of o-, m-, p-isomers)	1.495	ethyl iodide	1.92
p-xylene	1.500	bromoform	2.88

The quartz glass prisms of the cell are bonded to each other by a heat-resisting cement. Before the specimen of the glass to be measured is placed into the cell, it

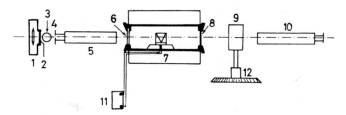

Fig. 79 Diagram of the apparatus for measuring refractive index at temperatures above the softening point of glasses.

is cut and optically polished to prismatic shape in order that full optical contact may be established between the surfaces of the specimen and those of the quartz glass cell.

A schematic diagram of the system of the cell (of refractive index n''), the inserted prism of the glass to be measured (refractive index n), and the passage of the monochromatic beam of rays through the system of prisms making up the cell can be found in Figure 80.

The source of light applied is a mercury vapour lamp 1, in front of which is placed an interference filter 2 ($\lambda = 546$ nm), or a monochromator. The radiation passes through the condenser 3, the slit of collimator 4 (which may be interchanged

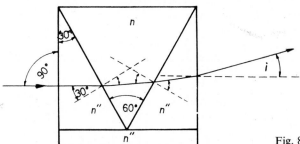

Fig. 80 Light passing through a V-block.

for an autocollimating attachment setting the incidence of the radiation perpendicular to the front surface of the cell), and collimator 5. Openings 6 and 8 equipped with optical quartz glass shutters close the tubular electric oven. Levelling of electric oven 7 is carried out by special screws. On emerging from the cell placed in the oven, the monochromatic ray of light passes through the Pekhanov prism 9 into the telescope of the spectrometer 10, where the deflection of the ray from its original direction is recorded. The refractive index of the glass is then calculated from the equation

$$n_\lambda = \left\{ \frac{1}{3} [2 \sin^2 i + \sqrt{3} n'' \sin i + 2n''^2 + \right.$$
$$\left. + (2\sqrt{3} \sin i + n'') \sqrt{n''^2 - \sin^2 i}] \right\}^{1/2}, \tag{156}$$

where n_λ is the refractive index of the specimen,

$\qquad i$ is the deflection of the rays from their original direction during their passage through the cell,

and $\quad n''$ is the refractive index of quartz glass for the given wavelength and temperature of the measurement.

The method can be applied up to temperatures above the softening point of quartz glass, but below its transformation range. The accuracy of the measurement depends upon the experimental arrangement of the measuring equipment and upon the accuracy of the reading of angle i.

114

Measuring the refractive index of glasses up to the softening point temperature.

Measurement can be made if the glass to be measured has been cut to a prism as shown in Figure 81.

If 2α is the angle of the prism, and 2β the recorded angle of the split ray on its emergence from the prism, the refractive index is computed from equation

$$n_\lambda = \frac{1}{\cos 3\alpha}(\sin^2\beta + 2\sin\beta\cos 2\alpha\sin 3\alpha + \cos^2\alpha)^{1/2}. \qquad (157)$$

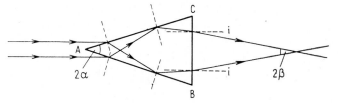

Fig. 81 Measurement of the refractive index up to the softening point of glasses.

A prism of the glass to be measured is again placed in the electric oven and the refractive index is gradually recorded for different temperatures. The upper limit of the temperature is set by the deformation of the prism of the glass under measurement. This then is the procedure for measuring the dependence of the refractive index upon the temperature in quartz glass. (See Figure 49).

Measuring the refractive index of glasses in the infra-red region of the spectrum

The high accuracy and general availability makes spectrometric methods most suitable for the measurement of the refractive indices of glasses in the infra-red region of the spectrum (of glasses transmissive to wavelengths of the infra-red region of the spectrum) [33].

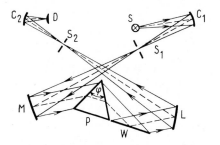

Fig. 82 Measurement of the index of refraction of glass in the infra-red region of the spectrum after Wadsworth: s — source of light; $c_{1,2}$ — focusing mirrors, S_1 — entrance slit; M — collimating mirror, P — prism; W — Wadsworth's mirror; L — lens; S_2 — exit slit; D — detector.

In view of the region of the measurement, the systems of lenses with satisfactory correction for aberrations are not easily optimized. That is why systems of mirrors [33] are usually resorted to. Use can be made of various types of arrangement, e.g., Wadsworth's, Littrow's, Littrow-Wadsworth's, etc. Figure 82 presents a schematic diagram of the Wadsworth arrangement.

Use is made of a number of discharge lamps as sources of monochromatic radiation (see Section 3.8). The sources of continuous radiation for the near infra-red region are high temperature radiators with tungsten and molybdenum filaments, for the medium infra-red region the Nernst glower and Globar (up to 10 μm), and for the range above 10 μm the Auer lamp.

The separation of a narrow spectral band from continuous spectrum sources can be accomplished with the help of a monochromator or special narrow-band filters.

Detection can be achieved by thermo-couples (the sensitivity for the more current combinations of metals ranging between 20 μV/μW), bolometers: metal-type bolometers (the sensitivity of which is about 0.9 μV/μW), thermistor-type bolometers (with sensitivity ranging between 0.7 and 1.2 μV/μW), and pneumatic bolometers (based on the principle of gas thermometers). Use can also be made of photoelectric cells, particularly for the near infra-red region, and of luminescent detectors.

Using optical methods to measure the homogeneity of glasses,

The methods applied in the measurement of the homogeneity of glasses are dealt with in considerable detail in reference [19]. We will thus confine ourselves to mentioning briefly only the optical methods making use of the difference between the refractive index of the inhomogeneity and the index of refraction of the surrounding glass.

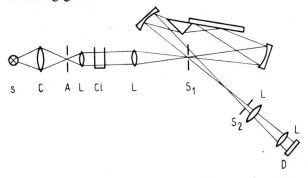

Fig. 83 Examination of inhomogeneities in glasses by Shelyubski's method: s — source of light; C — condenser; A — aperture; L — lens; Cl — cell; S_1, S_2 — entrance and exit slits of monochromator; D — detector.

One of the most widely-known methods of this type is *Shelyubski's method* described in detail in reference [21]. Crushed glass is placed in a cell filled with an immersion liquid, which is illuminated by a beam of light. Since the refractive index of the inhomogeneities differs from the index of refraction of the glass, the intensity of the transmitted light depends upon the temperature of the immersion liquid measured, and conclusion can be drawn as to the concentration and character of the inhomogeneities. If the glass, which is placed in the cell in crushed form, is fully homogeneous, a uniform index of refraction will be recorded at a certain temperature of the immersion liquid, and the maximum will exhibit abrupt sharpness. If, however, the glass is non-homogeneous (showing, for example, striae), the shape of the curve will be different: the maximum will be considerably lower and wider. The shape of the curve will thus characterize inhomogeneities present in the glass. For the diagram of the method the reader is referred to Figure 83.

Vidro, Khorolski, and Mironenko [22] also make use of crushed glass and an immersion liquid. The measurement is, however, carried out at room temperature. In principle, their measurement concentrates on the curve of spectral transmission in the visible region of the spectrum. If the glass is homogeneous, the transmission curve will exhibit a peak with a sharp maximum at a certain wavelength. If, on the other hand, the glass is non-homogeneous, the height of this maximum for the given wavelength will be reduced and the curve will manifest an enlarged maximum corresponding to the degree of the non-homogeneity.

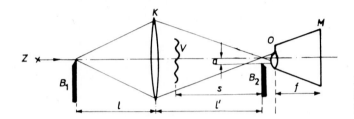

Fig. 84 Examination of inhomogeneities in glasses by Toepler's method.

One of the simplest methods of finding inhomogeneities in glass is by *screen projection*. The image can then be recorded on a diapositive photographic plate. For rough surfaces, use is made of immersion liquids. Enlarged photographs can also be made on contrastive photographic paper.

The *lattice method* [19] makes use of a line raster. The distortion of the transmitted rays caused by the inhomogeneities is recorded or photographed with the glass specimen placed in front of a line raster lattice of black and white stripes.

Toepler's method of penumbral arrangement has found particularly wide application in optical glass checks [54]. The arrangement of the apparatus for a convergent beam of rays is schematically sketched in Figure 84.

The incoming light, sharply defined by the edge, meets the condenser, which converts the divergent beam into a convergent one. The defining edge is imaged in the place where the other edge allows only a very narrow beam of the rays to pass. The image of the specimen is then displayed on the screen and, if the specimen is homogeneous, the whole picture is uniformly penumbrally illuminated. Adjusting the position of the second edge makes it possible to control the intensity of the illumination. If there is a stria in the glass, or if the surface of the glass manifests some irregularities causing the light to deflect, the image of the inlet edge will move either upwards or downwards. If it undergoes an upward shift, the image of the place where the stria is located will manifest a conspicuously higher luminos-

117

ity, and vice versa. Proper adjustment of the second edge will ensure maximum sensitivity.

Applying the principle of this method, Dubský and Fanderlik [55] have designed a simple apparatus, which can be employed in detecting and photographing inhomogeneities in glass. The apparatus consists of readily available component parts, e.g., an enlarger, a microscopic lamp, and a photographic camera. A schematic optical diagram is presented in Figure 85.

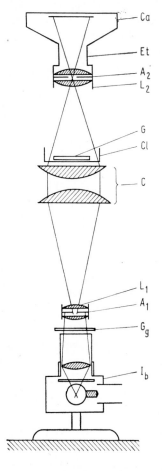

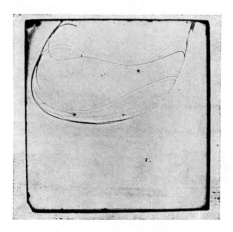

Fig. 86 Striae in lead crystal glass (not immersed).

◄ Fig. 85 Diagram of apparatus for photographing inhomogeneities in glass by Toepler's modified method: Ca — photographic camera; Et — extension tube, A_2 — aperture of camera; L_2 — lens of camera; Cl — cell; G — specimen of glass; C — condenser; L_1 — lens of enlarger; A_1 — aperture of enlarger; G_g — ground glass; I_b — illuminating bulb;

The specimen of the glass to be examined is placed, either free or in a cell filled with immersion liquid, in that part of the enlager where correction filters are normally inserted (above the condenser). The results obtained with the help of this apparatus are presented in Figures 86–89.

Schardin's method [56] is based on the observation of the deflection in one direction (the beam of rays being deflected in the shape of a narrow band), or in all the possible directions (with the beam of rays emitted by a light source passing

118

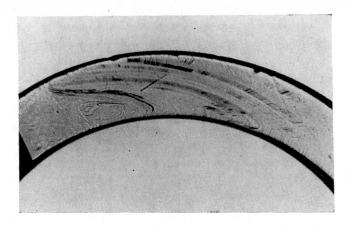

Fig. 87 Striae in a section of Simax glass tube (not immersed).

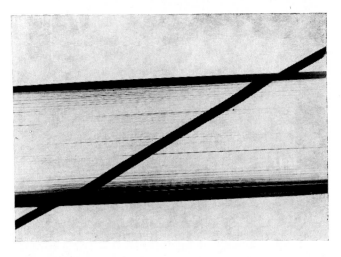

Fig. 88 Capillaries in the walls of quartz glass tube (immersed).

Fig. 89 Surface irregularities of quartz glass tube (not immersed).

through a circular aperture). A transparent screen with coloured stripes, or, in the latter case, with coloured annuli, is fixed at the place of the light source image and the glass specimen under examination is placed between the condensers. The displacement of the image of the light source caused by the inhomogeneities in the glass specimen or by the irregularities of the surface is defined by the angle of deflection. A certain colour corresponds to a certain angle of deflection and, consequently, to a certain degree of inhomogeneity. For the circular colour filter the recommendation is to arrange the colours in the following order starting from the centre: white, black, red, black, green, black, yellow, black, blue, black.

Inhomogeneities in glasses can also be examined by a number of other optical methods. The reader will find descriptions of some of them in [19, 20, 21, 22, 54, 56] of the bibliography.

4.2.8   Calculations

In calculating the optical properties of glasses, viz. the index of refraction, mean dispersion, and the Abbe value, as a function of composition, use is made of the additive factors determined for a wide range of the different types of the composition of glasses. The additive method of computing the properties of glasses is based upon the assumption that in a certain compositional region the value representing the property is equal to the sum of the values of the individual components in the glass, viz. oxides, silicates, or the regions of structural arrangement at shorter distances.

Calculation after L. I. Demkina

This method is one of the most precise methods ever developed to calculate the optical properties of glasses. It is applicable in the calculations of the optical properties of virtually all the currently produced optical glasses, i.e., the majority of the commercially marketed glasses. Demkina [16] derived the additive relationship for the calculation of the refractive index, mean dispersion, and the Abbe value:

$$K = \frac{\dfrac{a_1}{S_1} k_1 + \dfrac{a_2}{S_2} k_2 + \ldots + \dfrac{a_i}{S_i} k_i}{\dfrac{a_1}{S_1} + \dfrac{a_2}{S_2} + \ldots + \dfrac{a_i}{S_i}}, \tag{158}$$

where K is the calculated value of the refractive index, mean dispersion, and the Abbe value,

$a_1, a_2, a_3, \ldots, a_i$ the wt. percentages of oxides in the glass,

$S_1, S_2, S_3, \ldots, S_i$ the structural factors for the individual oxides, viz the coefficients giving the conversion of the mass units to units of volume, for which linearity remains valid in a wide range of the composition,

120

$k_1, k_2, k_3, \ldots, k_i$ the factors for the index of refraction, mean dispersion, and the Abbe value of the individual oxides.

The factors required for the numerical evaluation of equation (158) are given in Table 20.

Table 20

Oxide	S_i	k_{nD}	$k_{n_F-n_C} \cdot 10^5$	k_{v_d}
SiO$_2$-I	60	1.475	695	68
SiO$_2$-II	60	1.458	678	68
B$_2$O$_3$ BO$_4$	43	1.610	750	81
B$_2$O$_3$ BO$_3$	70	1.464	670	69
Al$_2$O$_3$	59	1.490	850	58
As$_2$O$_3$	107	1.570	1 600	36
Sb$_2$O$_3$	154	1.980	3 800	26
MgO	140	1.640	1 300	49
CaO	86	1.830	1 750	47
BaO	213	2.030	2 280	45
ZnO	223	1.960	2 850	34
PbO-I	343	2.460	7 700	19
PbO-II	223	2.460	7 700	19
PbO-III	223	2.500	11 600	13
Na$_2$O	62	1.590	1 400	42
K$_2$O	94	1.580	1 200	48

In Table 20 the values for boron oxide (B$_2$O$_3$), silicon dioxide (SiO$_2$), and lead oxide (PbO) are quoted for more than one coordination. In order to select the most suitable of these factors, one must observe several rules.

The simplest calculation is the one for the SiO$_2$ factors. For the majority of glasses the SiO$_2$ − I factor can be decided upon. The SiO$_2$ − II values can only be applied for glasses with a content of silicon dioxide exceeding 80 weight per cent.

The following rule concerns multi-component glasses, the composition of which puts them into a region bounded by

$$\Delta SiO_2 = n_{SiO_2} - [0.5(n_{ZnO} + n_{MgO}) + (n_{BaO} + n_{CaO}) +$$
$$+ 2(n_{B_2O_3} + n_{PbO} + n_{Na_2O}) + 4n_{K_2O}] \geq 0 \tag{159}$$

on the one hand, and by

$$\Delta SiO_2 = 0.30 \frac{n_{Na_2O}}{n_{Na_2O} + n_{K_2O}} + 0.09 \frac{n_{K_2O}}{n_{Na_2O} + n_{K_2O}}, \tag{160}$$

where n is the mol. percentage, on the other.

If the composition of the glass is within the region bounded by equations (159) and (160), factors $SiO_2 - I$, $B_2O_3 - I$, and $PbO - I$ will be used in the calculation.

For B_2O_3, which may be present in glass in coordinations $[BO_4]$ and $[BO_3]$, Demkina has established that coordination depends upon the so-called oxygen number O.

$$O = \frac{n_{PbO} + n_{BaO} + n_{CaO} + n_{K_2O} + n_{Na_2O} - n_{Al_2O_3}}{n_{B_2O_3}}. \qquad (161)$$

If the oxygen number of the glass, O, is greater than, or equal to, 1.2, and if ΔSiO_2 according to (159) is greater than 0, all the boron oxide in the glass will be in the $[BO_4]$ form.

If the oxygen number of the glass, O, is less than 1.2, the following relationship will be applied in the computation of the coordination of boron:

$$b_4 = O - 0.20, \qquad (162)$$

where b_4 denotes the ratio of BO_4 to BO_3.

If the oxygen number, O, is greater than 1.2, but ΔSiO_2 is less than zero, boron oxide will be present in the glass as both $[BO_4]$ and $[BO_3]$, the ratio being $2/3 : 1/3$.

Table 21

Group of glasses	$\xi = \dfrac{n_{PbO}}{n_{SiO_2} + n_{PbO}}$	Relationship for calculation		
		ξ_I	ξ_{II}	ξ_{III}
A	0—0.21		0	0
B	0.21—0.40	$0.11 + 0.475\,\xi$	$\xi - (\xi_I + \xi_{III})$	$0.0165\,(\beta - 0.85)$

For calculations concerning glasses with a PbO content the applicability of the $PbO - I$ coefficient is limited by the content of PbO not exceeding $45-50$ weight per cent. If the content of PbO is higher, calculation is inevitable.

If

$$\beta = \frac{4n_{K_2O}}{n_{SiO_2} - 2n_{PbO-I}}, \qquad (163)$$

glasses in the $K_2O - PbO - SiO_2$ system can be classified into two groups, A and B, according to the molar content of PbO, with the help of the relationship

$$\xi = \frac{n_{PbO}}{n_{PbO} + n_{SiO_2}}. \qquad (164)$$

Denoting $PbO - I$, $PbO - II$, and $PbO - III$ by ξI, ξII, and ξIII, respectively (if $\xi I + \xi II + \xi III = 1$), the contents of PbO – I, PbO – II, and PbO – III can be computed from the relationship listed in Table 21.

After calculating ξI, ξII, ξIII, we will obtain the values of n_{PbO-I}, n_{PbO-II}, and $n_{PbO-III}$ from equations

$$n_{PbO-I} = (n_{SiO_2} + n_{PbO})\,\xi_I,$$
$$n_{PbO-II} = (n_{SiO_2} + n_{PbO})\,\xi_{II}, \tag{165}$$
$$n_{PbO-III} = (n_{SiO_2} + n_{PbO})\,\xi_{III}.$$

For instance, for the glass of the composition

$$SiO_2 \quad 0.6237 \text{ mol. per cent}$$
$$PbO \quad 0.3363 \text{ mol. per cent}$$
$$K_2O \quad 0.0400 \text{ mol. per cent}$$

we get

$$\xi = \frac{n_{PbO}}{n_{PbO} + n_{SiO_2}} = \frac{0.3363}{0.9600} = 0.3503; \tag{166}$$

ξ is within the limits of 0.21 and 0.40, which means that the glass is in group B. Then

$$\xi I = 0.1100 + 0.475\xi = 0.1100 + 0.475 . 0.3503 = 0.2764,$$
$$n_{PbO-I} = (n_{SiO_2} + n_{PbO})\,\xi I = 0.9600 . 0.2764 = 0.2653,$$
$$\beta = \frac{4n_{K_2O}}{n_{SiO_2} - 2n_{PbO-I}} = \frac{4 . 0.0400}{0.6237 - 2 . 0.2653} = 1.72, \tag{167}$$
$$\xi III = 0.0165\,(\beta - 0.85) = 0.0165\,(1.72 - 0.85) = 0.0144,$$
$$n_{PbO-III} = \xi III\,(n_{SiO_2} + n_{PbO}) = 0.0144 . 0.9600 = 0.0138,$$
$$n_{PbO-II} = n_{PbO} - (n_{PbO-I} - n_{PbO-III}) = 0.3363 - (0.2653 - 0.0138) = 0.0572.$$

The total content of PbO must thus be divided as follows:

PbO – I : 0.2653 mol. per cent, i.e.,
0.5083 weight per cent,

PbO – II : 0.0572 mol. per cent, i.e.,
0.1096 weight per cent,

PbO – III: 0.0138 mol. per cent, i.e.,
0.0264 weight per cent.

Example:

Let us calculate the refractive index and mean dispersion of the glass the composition of which is quoted in Table 22.

The computation proceeds along the lines indicated in Table 22.

For the refractive index we thus get

$$n_d = \frac{1.570\,05}{0.965\,84} = 1.6256 \text{ (meas. 1.6259)},$$

123

and for mean dispersion

$$n_F - n_C = \frac{1690.36 \cdot 10^{-5}}{0.965\,84} = 0.017\,50 \ (\text{meas. } 0.017\,56).$$

In the calculation use was made of factors $SiO_2 - I$ and $PbO - I$.

Table 22

Component	a_i [wt. %]	$\dfrac{a_i}{S_i}$	$\dfrac{a_i}{S_i} \cdot k_{n_i}$	$\dfrac{a_i}{S_i} \cdot k_{n_F - n_C)_i}$
SiO_2	45.5	0.7583 3	1.118 54	527.04
PbO	48.1	0.1402 1	0.344 97	1 079.84
K_2O	6.0	0.0638 3	0.100 85	76.59
As_2O_3	0.3	0.0028 0	0.004 40	4.48
Sb_2O_3	0.1	0.0006 5	0.001 28	2.46
Σ	100.0	0.965 82	1.570 04	1 690.41

Calculation after Appen

The calculation proceeds from the additive relationship [57, 24], viz.

$$k = \frac{\Sigma m_i \bar{k}_i}{\Sigma m_i}, \tag{168}$$

where K is the calculated value of the refractive index, mean dispersion,

$\bar{k}_i$ the mean value of the property of each of the components (factors),

and m_i the content of each component of the glass in mol. per cent.

Table 23 lists the mean values of the properties, $\bar{k}_i$ (refractive index, mean dispersion), of each component of the glass.

From this table it follows that for silicon oxide, boron trioxide, lead oxide, cadmium oxide, and titanium dioxide, the $\bar{k}_i$ values dependent upon the composition of the glass must be computed from the following empirical equations. For silicon oxide:

If $m_{SiO_2} < 67$ mol. per cent, then $n_{SiO_2} = 1.475$,
$(n_F - n_C)_{SiO_2} = 675 \cdot 10^{-5}$;

If $m_{SiO_2} > 67$ mol. per cent, then $n_{SiO_2} = 1.475 - 0.0005\,(m_{SiO_2} - 67)$,
$(n_F - n_C)_{SiO_2} = 675 \cdot 10^{-5}$. $\qquad (169)$

For boron trioxide:

Computing the B_2O_3 factor, we must first calculate the value of ψ_B from equation

$$\psi_B = \frac{m_{Na_2O} + m_{K_2O} + m_{BaO}}{m_{B_2O_3}}. \tag{170}$$

From Table 24 we then find the factor by solving numerically the equation presented in the table.

Table 23

Oxide	k_{n_D}	$k_{n_F - n_C} \cdot 10^5$
SiO_2	1.4585—1.475	675
TiO_2	2.08—2.23	5 000—6 200
ZrO_2	2.17	—
B_2O_3	1.460—1.710	640—900
Al_2O_3	1.520 (1.70)	850
Sb_2O_3	2.55	7 700
BeO	1.595	890
MgO	1.610 (1.57)	1 110
CaO	1.730	1 480
SrO	1.770	1 630
BaO	1.880	1 890
ZnO	1.710	1 650
CdO	1.805—1.925	2 270—2 930
PbO	2.150—2.350	5 280—7 440
Li_2O	1.695 (1.655)	1 380 (1 300)
Na_2O	1.590 (1.575)	1 420 (1 400)
K_2O	1.575 (1.595)	1 300 (1 320)
Cs_2O	1.700 (1.74—1.8)	—

Table 24

Content of SiO_2 [mol. %]	Numerical value of ψ	Equation for numerical value of $\bar{n}_{B_2O_3}$, $\Delta\bar{n}_{B_2O_3}$
44—64	$\psi > 4$	$\bar{n}_{B_2O_3} = 1.710$
	$4 > \psi > 1$	$\bar{n}_{B_2O_3} = 1.710 - 0.048 \, (4 - \psi)$
	$1 > \psi > 1/3$	$\bar{n}_{B_2O_3} = 1.470 + 0.048 \, (3 - 1/\psi)$
	$\psi < 1/3$	$\bar{n}_{B_2O_3} = 1.470$
71—80	$\psi > 1.6$	$\bar{n}_{B_2O_3} = 1.710$
	$1.6 > \psi > 1.0$	$\bar{n}_{B_2O_3} = 1.710 - 0.12 \, (1.6 - \psi)$
	$1 > \psi > 1/2$	$\bar{n}_{B_2O_3} = 1.520 + 0.12 \, (2 - 1/\psi)$
	$1/2 > \psi > 1/3$	$\bar{n}_{B_2O_3} = 1.470 + 0.048 \, (3 - 1/\psi)$
	$\psi < 1/3$	$\bar{n}_{B_2O_3} = 1.470$
44—80	$\psi > 4$	$\Delta\bar{n}_{B_2O_3} \cdot 10^5 = 900$
	$\psi < 4$	$\Delta\bar{n}_{B_2O_3} \cdot 10^5 = 900 - 65 \, (4 - \psi)$

For lead oxide:

$$n_{PbO} = 2.350 - 0.0067 \, (\Sigma \, m_{Me_mO_n} - 50), \qquad (171)$$
$$(n_F - n_C)_{PbO} \cdot 10^5 = 7440 - 72 \, (\Sigma \, m_{Me_mO_n} - 50),$$

125

where Me_mO_n corresponds to SiO_2, B_2O_3, and Al_2O_3 and

$$50\% < \Sigma m_{Me_mO_n} < 80\%.\qquad(172)$$

For cadmium oxide:

$$n_{CdO} = 1.925 - 0.004\,(\Sigma m_{Me_mO_n} - 50),$$
$$(n_F - n_C)_{CdO} \cdot 10^5 = 2930 - 22\,(\Sigma m_{Me_mO_n} - 50).\qquad(173)$$

For titanium dioxide:

$$n_{TiO} = 2.23 - 0.005\,(m_{SiO_2} - 50),$$
$$(n_F - n_C)_{TiO_2} \cdot 10^5 = 6200 - 40\,(m_{SiO_2} - 50).\qquad(174)$$

The calculation of the optical properties of glasses after Appen is dealt with in detail in [23]. The calculations of the optical properties of glasses after Huggins, Kuan-Han Sun and others are described in references [23, 24, 25].

References

[1] M. V. Volkenštejn, Struktura a fyzikální vlastnosti molekul (Structure and the Physical Properties of Molecules). NČSAV, Prague 1962.
[2] D. G. Holloway, The Physical Properties of Glass. Wykenham Publ. Ltd., London—Winchester 1973.
[3] J. Brož et al., Základy fyzikálních měření (Foundations of Physical Measurements) II B. SPN, Prague 1974.
[4] M. Fanderlik, Struktura skel (Structure of Glasses). SNTL—ALFA, Prague—Bratislava 1971.
[5] L. Prod'homme, A new approach to the thermal change in the refractive index of glasses. Phys. Chem. Glasses, 1 (1960) pp. 119—122.
[6] A. Winter, Transformation region of glass. J. Am. Ceram. Soc., 26 (1943) pp. 189—200.
[7] H. A. Mc Master, Variations of refractive index of glass with time and temperature in the annealing region. J. Am. Ceram. Soc., 28 (1945) pp. 1—4.
[8] W. P. Collyer, Study of time and temperature effects on Glass in the annealing range. J. Am. Ceram. Soc., 30 (1947) pp. 338—344.
[9] J. E. Danyushevskyi, Osnovy lineynovo otzhiga opticheskovo stekla (Foundations of Linear Annealing of Optical Glass). G. I. P., Moscow 1959.
[10] L. H. Adams and E. D. Williamson, The annealing of glass. J. Franklin Inst., 190 (1920) pp. 597—631, 835—870.
[11] N. M. Brandt, Annealing of 517.645 borosilicate optical glass. I. Refractive index. J. Am. Ceram. Soc., 34 (1951) No. 11, pp. 332—338.
[12] H. R. Lillie and H. N. Ritland, Fine annealing of optical glass. J. Am. Ceram. Soc., 37 (1954) No. 10, pp. 466—473.
[13] F. Schill et al., Chlazení skla a kontrola pnutí (Annealing of Glass and Control of Internal Stress). SNTL, Prague 1968.
[14] T. Izumitani, Report of the Osaka Industrial Research Institute, No. 3, 1958.
[15] K. Fajans and N. J. Kreidl, Stability of lead glasses and polarization of ions. J. Am. Ceram. Soc., 31 (1948) pp. 105—114.
[16] L. I. Demkina, Issledovanyie zavisimosti svojstv stekol ot ikh sostava (Study of the Dependence of the Properties of Glasses upon the Composition of Glasses). C.I.O.P., Moscow 1958.
[17] Zd. Horák, Fr. Krupka, and V. Šindelář, Technická fyzika (Technical Physics), SNTL, Prague 1961.

126

[18] A. Winter, Glass formation. *J. Am. Ceram. Soc.*, *40* (1957) pp. 54—58.

[19] M. Fanderlik, Vady skla (Defects of Glass). SNTL—SVTL, Prague 1963.

[20] V. I. Shelyubskyi, Novyi metod opredelenyia i kontrolya odnorodnosti stekla (A new method of determining and checking the homogeneity of glass). *Steklo i keramika, 17* (1960) No. 8, pp. 17—22.

[21] M. Cable, S. D. Walters, A test of Shelyubski's method for determining the homogeneity of glass. *Glass Technology 21* (1980) No. 6, pp. 279—283.

[22] L. I. Vidro, Yu. M. Khorolskyi, and L. A. Mironenko, Ustanovka dlya kontrolya stepeni odnorodnosti stekla (Apparatus determining the degree of homogeneity of glass). *Steklo i keramika, 17* (1960) No. 8, pp. 22—25.

[23] M. A. Matveyev, G. M. Matveyev, and B. H. Frenkel, Raschoty po khimii i tekhologyi stekla (Calculations of Chemical and Technological Properties of Glass). I.L.S., Moscow 1972.

[24] M. B. Volf, Sklářské tabulky a výpočty (Glass Maker's Tables and Calculations). Průmyslové vydavatelství, Prague 1952.

[25] M. Fanderlik, O. Přidal, and F. Trenz, Sklářské praktikum (Glass Maker's Practicum). Průmyslové vydavatelství, Prague 1951.

[26] H. M. Pavlushkin et al., Steklo-spravochnik (Glass Maker's Handbook). Stroyizdat, Moscow 1973.

[27] A. Winter, Les formateurs de reseaux vitreux et quelques propriétés optiques des verres. IVᵉ Congres International du Verre, Paris 1956, pp. 415—418.

[28] G. Rosenthal, Das Sichtbarmachen von Schlieren und Messung kleiner Unterschiede in Brechnungsindex mit Hilfe der Interferenz. *Glastechn. Ber., 23* (1950) p. 131.

[29] K. H. Sun and M. L. Huggins, The effect of chemical composition on the relationship between refractive index and Abbe value for binary systems. *J. Soc. Glass Technol., 29* (1945) pp. 192—196.

[30] V. Süsser and I. Fanderlik, Vliv ThO_2, La_2O_3, CdO, ZrO_2 na fyzikálně chemické vlastnosti optického skla (The effect of ThO_2, La_2O_3, CdO, ZrO_2 upon the physico-chemical properties of optical glass). *Sklář a keramik* (1963), No. 10, pp. 265—268.

[31] I. Fanderlik, Speciální optická skla s vysokým indexem lomu (Special optical glasses with a high index of refraction). *Věda a výzkum v průmyslu sklářském*, 1967, Series X, pp. 17—65.

[32] H. Scholze, Glass—Natur, Struktur und Eigenschaften. Springer Verlag, Berlin—Heidelberg, New York 1977.

[33] M. Prokop, Měření indexu lomu optických materiálů v infračervené oblasti (Measuring the refractive index of optical materials in the infra-red region). *Jemná mech. a opt., 3* (1958) No. 1, pp. 12—16; No. 2, pp. 47—48.

[34] B. Havelka, Geometrická optika (Geometrical Optics). ČSAV, Prague 1955.

[35] W. Gefcken, Mehrstoffsysteme zum Aufbau optischer Gläser. Teil I.: Grundsatzliche Beziehungen. *Glastechn. Ber., 34* (1961) No. 3, pp. 91—101.

[36] M. Faulstich, Mehrstoffsysteme zum Aufbau optischer Gläser. Teil II.: System B_2O_3—La_2O_3——ThO_2—Ta_2O_5—Nb_2O_5. *Glastechn. Ber. 34* (1961) No. 3, pp. 102—107.

[37] W. Jahn, Mehrstoffsysteme zum Aufbau optischer Glaser. Teil III.: Neue optische Gläser auf Fluoridbasis. *Glastechn. Ber., 34* (1961) pp. 107—120.

[38] F. Reitmayer, Die Dispersion einiger optischer Gläser im Spektralgebiet von 0.33 bis 2.0 μm und ihre Abhängigkeit von den Eigenfrequenzen der Gläser. *Glastechn. Ber., 34* (1961) No. 3, pp. 122—130.

[39] B. Meinecke, Ueber die Austauschbarkeit thoriumoxydhaltiger Gläser für die Optik. *Glas-Email-Ker.-Techn.*, (1959) No. 6, pp. 209—212.

[40] G. F. Brewster, J. F. Kunz, and S. L. Rood, Dispersion of some optical glasses in the visible and infrared regions. *J. Opt. Soc. Am. 48* (1958) pp. 534—536.

[41] N. Neuroth, Der Temperatureinfluss auf die optischen Konstanten von Glass im Gebiet starker Absorption. *Glastechn. Ber.*, *28* (1955) pp. 411—422.

[42] W. Vogel and K. Gerth, Ueber Modelsilikat-Gläser und ihre Konstitution. Die Glassysteme LiF—BeF$_2$, NaF—BeF$_2$, KF—BeF$_2$ und RbF—BeF$_2$. *Glastechn. Ber.*, *31* (1958) pp. 13—27.

[43] W. Vogel and K. Gerth, Zur Struktur von Fluoridgläser. Die Glassysteme MgF$_2$—BeF$_2$, CaF$_2$—BeF$_2$, SrF$_2$—BeF$_2$. *Silikattechn.*, *9* (1958) pp. 353—358.

[44] W. Vogel and K. Gerth, Zur Struktur von Fluoridgläser. Die ternaren Alkali-erdalkali-beryllium-fluoridglassysteme MgF$_2$, CaF$_2$, SrF$_2$, BaF—KF—BeF; MgF$_2$, CaF$_2$, SrF$_2$—NaF——BeF$_2$; MgF$_2$—LiF—BeF$_2$. *Silikattechn.*, *9* (1958) pp. 495—501.

[45] I. Gregora, M. Závětová, and V. Vorlíček, Refractive Index of Amorphous Arsenic. Amorphous Semiconductors 78, Pardubice, pp. 416—419.

[46] Z. Cimpl, V. Husa, F. Kosek, and M. Matyáš, Refractive Index of Glassy Arsenic Trisulphide. Amorphous Semiconductors 78, Pardubice, pp. 428—431.

[47] T. F. Mazec, S. K. Pavlov, M. Závětová and B. Vorlíček, Izmenenie koefitsienta pogloshchenia i pokazatelia prelomlenia v amorfnykh plenkakh 95 As$_2$Se$_3$. 5 As$_2$Te$_3$ pod deistvyem electricheskoskovo polya (Change in the Coefficient of Absorption and the Refractive Index of Amorphous Coatings 95 As$_2$Se$_3$. 5 As$_2$Te$_3$ under the Efect of the Electric Field). Amorphous Semiconductors 78, Pardubice, pp. 432—435.

[48] F. C. Everstejn, J. H. Stevels, and H. I. Waterman, The density, refractive index and specific refraction of vitreous boron oxide and of sodium borate glasses as functions of composition, method of preparation, and rate of cooling. *Phys. Chem. Glass*, *1* (1960) pp. 123—133.

[49] Vl. Novotný, Zpevňování skla (Reinforcement of Glass). SNTL, Prague 1972.

[50] D. Kocíková and J. Kocík, Speciální optická skla fluoridová (Special optical fluoride glasses). *Věda a výzkum v průmyslu sklářském*, Series VI, 1960.

[51] W. H. Armistead, USA Patent No. 2435995.

[52] I. Fanderlik and M. Skřivan, Utilisation de la méthode mathématico-statistique des experiences planifiés pour suivre la variation des propriétés physiques des verres des cristal en plomb en fonction de leur composition. *Verres et Réfr.*, *26* (1972) pp. 19—23.

[53] ČSN (Czechoslovak Standard) 71 0126 — Měření indexu lomu metodou Obreimovou (Measuring the Refractive Index by Obreim's Method).

[54] A. Toepler, Beobachtungen nach einer neuen Optischen Methode. Bonn, 1861.

[55] F. Dubský a I. Fanderlik, Fotografování nehomogenit ve sklech modifikovanou Teoplerovou polostínovou metodou v jednoduchém provedení (Photographing non-homogeneities in glasses by Toepler's modified half-shade method (simplified version)). *Sklář a keramik*, *28* (1978) No. 2, pp. 36—38.

[56] H. Schardin and G. Stamm, Prüfung von Flachglas mit Hilfe eines farbigen Schlierverfahrens. *Glastechn. Ber.*, *20* (1942) pp. 249—258.

[57] A. A. Appen, Khimia stekla (Chemistry of Glass). Khimia, Leningrad 1970.

[58] K. S. Evstropiev and A. O. Ivanov, Physikalisch-chemische Eigenschaften von Germanium-Gläsern. In: Advances in Glass Technology, Vol. 2. Plenum Press, New York 1963.

[59] A. A. Appen, Berechnung der optischen Eigenschaften der Dichte und des Asdehnungskoeffizienten von Silikatgläsern aus ihrer Zusammensetzung. *Ber. Akad. Wiss. UdSSR*, *69* (1949) pp. 841—844.

[60] H. Scholze, Deutung der unterschiedlichen Einflüsse des Wassergehaltes auf Lichtbrechung und Dichte von Gläsern verschiedlicher Zusammensetzung. *Glass-Email-Keramo-Techn.*, *19* (1968) pp. 389—390.

[61] M. L. Huggins and K. H. Sun, Calculation of density and optical constants of a Glass from its composition in weight percentage. *J. Am. Ceram. Soc.*, *26* (1943) pp. 4—11.

[62] F. Told, Systematik und Analyse der optischen Gläser hinsichtlich Brechungsvermögen und Dichte. *Glastechn. Ber.*, *33* (1960) pp. 303—304.

[63] R. M. Waxler and G. W. Clerk, The effect of temperature and pressure on the refractive index of some oxide glasses. *J. Res. Nat. Bur. Stand, 77a* (1973) pp. 755—763.

[64] L. I. Demkina, Fiziko-khimicheskye osnovy proizvodstva opticheskovo stekla (Physico-Chemical Foundations of the Production of Optical Glass). Khimia, Leningrad 1976.

[65] V. Novotný, Chlazení skla a kontrola pnutí (Annealing of Glass and Control of Internal Stress). SNTL, Prague 1968.

[66] I. Fanderlik and M. Skřivan, Koncentrační gradient v povrchové vrstvě způsobený těkáním složek skloviny (Concentration gradient in the surface layer caused by volatilization of the components of the melt). *Silikáty,* (1973) No. 4, pp. 295—302.

4.3   ABSORPTION OF RADIATION IN GLASSES

4.3.1   Introduction

Classical dispersion theory interprets absorption in terms of resonating electric dipoles present in the dielectric. When radiation passes through glass the electrons, or molecules, excited by the luminous (radiant) wave, start oscillating harmonically (see Section 4.2). The energy that they receive is proportional to the frequency of the incident radiation.

The largest vibrations are exhibited by the charges of the so-called dipoles in resonance; in that case glass absorbs above all the frequencies that correspond to the frequencies of the free oscillations of the charges in the dipoles.

4.3.2   Theory

The wave function of the atom defines all the energetic states of the atom that are governed by the transitions in the electron shells. The molecular wave function is more complex. The inner energy of the molecule, E_i, falls into three parts:

$$E_i = E_e + E_v + E_r + W, \tag{175}$$

where E_e is the energy of the electrons,

       E_v is the energy of the vibrational motion,

       E_r is the energy of the rotational motion,

and W is the energy of the interaction,

where

$$E_e \gg E_v \gg E_r. \tag{176}$$

An atom, or a molecule, can thus emit, or absorb, only photons the energy of which, hv, is equal to the difference between the energy levels. For a molecule, for example,

$$hv = E'' - E' = (E_e'' - E_e') + (E_v'' - E_v') + (E_r'' - E_r'). \tag{177}$$

We thus have electronic spectra, vibration spectra, rotation spectra, and combinations of the three. Electron transitions are characterized by radiation or absorption in the visible and the ultra-violet regions of the spectrum, the vibrational transitions in the near infra-red region, and the rotational transition in the far infra-red region of the spectrum and in the region of longer waves. We shall, of course, be interested in the electronic spectra of the atom, or molecule, arising from transitions of the electrons from one energy level to another as a result of external perturbation. The charge density around the nuclei may be deformed and polarized by the neighbouring ionic, atomic, or molecular energy fields. Hence, glasses do not manifest absorption at one single wavelength only but over a range of energies.

According to the configuration of the electrons round the nucleus, all atoms, or ions, can roughly be classified into the following groups [1]:

(a) atoms the orbits of which are all filled by electrons. These atoms do not form compounds, neither do they cause colouration (absorption bands). The group includes the rare gases: helium, neon, argon, krypton, and xenon;

(b) atoms whose outer shells are incomplete. They form compounds, their salts, however, being colourless: barium, strontium, etc.;

(c) atoms with two incomplete shells. This group includes transition elements like nickel, manganese, iron, cobalt, chromium etc. Their ions are coloured and they manifest absorption in the optical region of the spectrum;

(d) atoms three shells of which are incompletely filled. Transitions of electrons can occur in the inner orbits the energies of which are not effected by the forces due to neighbouring atoms, and absorption covers a very narrow range of energies only. The group includes the rare earths.

As we shall discuss below in more detail, colourless silicate glasses, for instance, absorb radiation of the wavelengths shorter than $160-300$ nm (viz. that of the ultra-violet region of the spectrum) and also radiation at wavelengths longer than 5 μm (viz. that of the infra-red region of the spectrum). In conventional silicate glasses the position of the absorption edge in the ultra-violet and the infra-red regions of the spectrum depends upon the structure of the glass. Only after absorbing colour centres, or, according to Weyl [6, 8], "inorganic chromophores", have been introduced, will these glasses acquire colour.

Since the ions of the transition elements are the most widespread colouring components of glasses, we shall in the following section, deal with the absorption spectra from the point of view of the *theory of the ligand fields,* which has recently, been a major contribution to the elucidation of the problem of the valency states of ionic colorants in glasses. For more theoretical information on this subject see Bates [2], Wong, Angell [34] and others; the application of the theory of the ligand fields to the generation of colour in glasses is dealt with by Bamford [3] and others. Our discussion of the problem will be based on Bates [2].

One of the basic difficulties in the theoretical study of the absorption spectra of the ions of the transition metals and the rare earths in glasses is the fact that these ions are present in different valencies and coordinations. They are characterized by a partly filled inner electron shell: the *d*-orbit of the ionic electrons of the transition metals has 1 to 9 out of ten potential electrons, the *f*-orbit of the electrons of the rare earths 1 to 13 out of fourteen potential electrons. The majority of the studies concerning this problem have therefore concentrated upon the valencies or the coordinations of such ions in glasses and upon the corresponding absorption bands as compared with their known states in crystals or solutions.

Ligand field theory

This theory assumes that an ion of the transition element is the central ion of the complex and is thus surrounded by atoms, or molecules, the so-called ligands. The absorption spectra of these complex structures can be classified according to the intensity of the absorption bands into two groups, as shown in Figure 90.

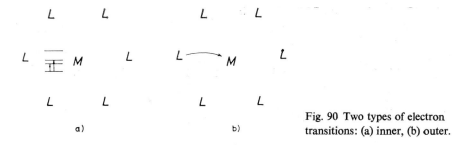

Fig. 90 Two types of electron transitions: (a) inner, (b) outer.

The *weak absorption bands* (Figure 90a) are produced by the internal transitions between the *d*-electron levels of the central ion modified by the ligand field. The absorption bands corresponding to the quoted transitions occur in the near infra-red, visible, and near ultra-violet regions of the spectrum. Despite the fact that the $d^n - d^n$ transitions of the electrons (*n* denoting the number of the *d*-level electrons) are forbidden by Laporte's rule, the electric dipole transitions coupled with molecular vibrations relax this rule thus making the electronic transitions possible.

The *strong absorption bands* arise from electronic transitions, which can be viewed roughly as a transfer of the charge from the ligands to the central ion or from the central ion to the ligands, as represented schematically in Figure 90b. Since Laporte's rule does not forbid them, their intensity is $100-1000$ times higher than with the internal transitions. These absorption bands occur mainly in the ultra-violet region, sometimes also in the short-wave visible region of the spectrum, where they then obscure the absorption bands associated with the internal transitions.

The strong absorption bands also arise from transitions of the electrons between two ions of the same element in different valency states, e.g., $Fe^{2+} \rightarrow Fe^{3+}$.

The ligand field theory postulates splitting of the energy levels of the partially filled shells of the free ion due to the effect of the energy fields of the ligands that

131

surround them, the basic symmetry of these complexes of transition metals being octahedral, in a number of cases even tetrahedral.

Let us first consider the effect of the ligand fields on a single d^1-electron, as, for example, in Ti^{3+}. In a free ion this electron will occupy any one of the five

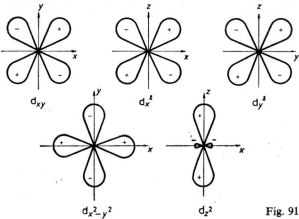

d_{xy} d_x^2 d_y^2

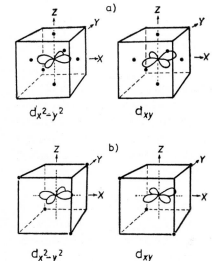

$d_{x^2-y^2}$ d_{z^2} Fig. 91 d-orbitals of electrons.

$d_{x^2-y^2}$ d_{xy}

a)

b)

$d_{x^2-y^2}$ d_{xy} Fig. 92 $d_{x^2-y^2}$ orbitals and d_{xy} orbitals in an octahedral (a) and tetrahedral (b) ligand field.

d-orbitals which exhibit degeneracy, their energies being equal. In conformity with the Schrödinger wave equation the orbitals with octahedral or tetrahedral symmetry denoted as d_{xy}, d_{xz}, d_{yz}, $d_{x^2-y^2}$, and d_{z^2} can be considered the most probable (see Figure 91).

Thus, if cation is the centre of an octahedral complex, viz. with ligands at $\pm x$, $\pm y$, and $\pm z$, the d-orbit does no longer manifest the same energy. In Figure 92 are diagrams of the $d_{x^2-y^2}$ and d_{xy} orbitals of an ion in an octahedral environment.

132

In view of the fact that the ligands are negatively charged, the electrons will tend to avoid the $d_{x^2-y^2}$ orbital, the lobes of which lie in the direction between the ligands. The effect of an octahedral arrangement thus consists in the removal of the degeneracy of the five d-orbitals, so that they undergo re-grouping into three (d_{xy}, d_{xz}, d_{yz}) γ_5 — orbitals (the lower group) and two $(d_{x^2-y^2}, d_z^2)$ γ_3-orbitals (the upper group). See Figure 93.

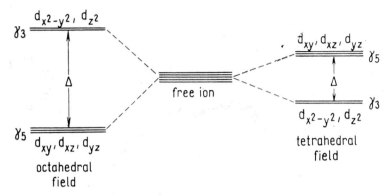

Fig. 93 Energy levels of the d-orbitals in octahedral and tetrahedral fields.

Consider now the ion to be in a tetrahedral complex. In Figure 92b the $d_{x^2-y^2}$ and d_{xy} orbitals of the central ion are shown, this time in a tetrahedral environment. In this case the situation is the reverse of that for octahedral symmetry. The lobes of the $d_{x^2-y^2}$ orbital now lie in the direction between the ligands, while the lobes of the d_{xy} orbital, although not pointing directly towards the ligands, lie closer to them. Thus the d_{xy} orbital is destabilized with respect to the $d_{x^2-y^2}$ orbital. Again by symmetry the d_{xy}, d_{xz}, and d_{yz} orbitals are degenerate, and calculations show that the d_{z^2} is degenerate with the $d_{x^2-y^2}$ orbital.

With tetrahedral coordination, in the same way as with octahedral, five d-orbitals are split into two groups, viz. $\gamma_5(d_{xy}, d_{xz}, d_{yz})$ and $\gamma_3(d_{x^2-y^2})$; the order of the levels however is reversed here. See Figure 93.

The energy difference between the γ_3 and γ_5 electrons, denoted as Δ, can be ascertained from the position of the absorption band.

From the theory of spatial symmetry it can further be established that Δ (tetrahedral) $= -\dfrac{4}{9} \Delta$ (octahedral) for the same ligands and the same ligand to central ion separation. Under the influence of ligand fields of lower symmetry the d-orbitals are split again, as illustrated in Figure 94 for a tetrahedral plane complex (i.e., with ligands at $\pm x$, $\pm y$).

This is however a special case, which will not be considered further. We have so far confined ourselves to the states and energy levels of a single electron. But if an ion has several d^1-electrons, the levels are not degenerate in a free ion any more, due to the effect of the interaction between the electrons. These electron states require some kind of annotation: l will denote the angular momentum of a single electron (s the spin momentum), and **L** the total angular momentum of a single ion (**S** the total spin momentum):

$$l = 0 \quad 1 \quad 2 \quad 3$$
$$\quad\quad s \quad p \quad d \quad f,$$

and for $L = 0 \quad 1 \quad 2 \quad 3 \quad 4 \quad 5 \quad 6$
$$\quad\quad\quad\quad S \quad P \quad D \quad F \quad G \quad H \quad I.$$

For the octahedral d^1-system, thus e.g. $[Ti\,(H_2O)_6]^{3+}$, the Ti^{3+} ion in the free state has a single ^2D level. The superscript refers to spin degeneracy $(2S + 1)$,

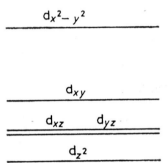

Fig. 94 Energy levels of the d-orbitals of electrons in a square planar field.

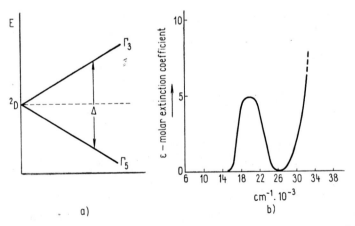

a)

b)

Fig. 95 Energy level diagram for d^1-systems (a); absorption spectrum of $[Ti(H_2O)_6]^{3+}$ (b).

which corresponds to two possible spin directions of the electron. Under the action of the ligands the level will split into an upper, Γ_3, and lower, Γ_5, level, as shown in Figure 95a.

134

The Γ_5 state will clearly be the ground state of the ion here, and $\Gamma_5 \to \Gamma_3$ the only possible transition of the electrons giving rise to a single absorption band (Figure 95b). From the position of the absorption band it is evident that the value of Δ for $[\mathrm{Ti(H_2O)_6}]^{3+}$ equals 20 000 $[\mathrm{cm}^{-1}]$.

Let us now consider an ion of a transition metal with several d-electrons. This case is by far more complex than the one of an ion with a single d^1-electron, because owing to the coupling between the electrons many of these states will have different energies. Each of the levels with nearly the same energy (multiplet terms) can be characterized by a particular value of $\mathbf{L}$ and $\mathbf{S}$ (Russel-Saunders coupling). This type of coupling only breaks down when the spin-orbit coupling becomes large. Figure 96 gives the free ion $\mathbf{L}$, $\mathbf{S}$-levels for the d^3 configuration.

Fig. 96 Free ion levels of the d^3-configuration

According to Slater, Condon, and Shortley [35], the energy differences between these terms can be evaluated as multiples of certain radial integrals of electrostatic interaction denoted by F_0, F_2, and F_4; Racah [36] expressed them in terms of three parameters, A, B, and C:

$$A = F_0 - 49F_4,$$
$$B = F_2 - 5F_4,$$
$$C = 35F_4.$$

Table 25 lists the lower free ion term spacings expressed in Racah's parameters B, C, if the energy of the ground state is considered to be equal to zero.

For all free ions of the transition metals it has been established that the ratio C/B is almost constant and $\cong 4.5$.

If an ion containing several d-electrons is exposed to the effect of an octahedral ligand field, the levels of the electrons will be determined both by the interactions between the d-electrons and by the effect of the ligands. These levels are characterized by the atomic quantum numbers $^{2S+1}\Gamma$ (e.g., $^3\mathbf{F}$, $^1\mathbf{D}$, $^3\mathbf{P}$ etc.). Bethe [37] has classified the levels in an octahedral field in terms of the corresponding free ion states as:

135

Free ion state	O_h Ligand field states
$S \rightarrow$	Γ_1
$P \rightarrow$	Γ_4
$D \rightarrow$	$\Gamma_3 + \Gamma_5$
$F \rightarrow$	$\Gamma_2 + \Gamma_4 + \Gamma_5$
$G \rightarrow$	$\Gamma_1 + \Gamma_3 + \Gamma_4 + \Gamma_5$
$H \rightarrow$	$\Gamma_3 + 2\Gamma_4 + \Gamma_5$

Table 25

Configuration	State	Energy
$d^1 \equiv d^9$	^2D	
$d^2 \equiv d^8$	^3F	0
	^1D	$5B + 2C$
	^3P	$15B$
	^1G	$12B + 2C$
	^1S	$22B + 7C$
$d^3 \equiv d^7$	^4F	0
	^4P	$15B$
	^2G	$4B + 3C$
	^2H	$9B + 3C$
	^2P	$9B + 3C$
	....	.....
$d^4 \equiv d^6$	^5D	0
	^3H	$4B + 4C$
	^3_aP	$16B + 5\frac{1}{2}C - \frac{1}{2}(912B^2 - 24BC + 9C^2)^{\frac{1}{2}}$
	^3_aF	$16B + 5\frac{1}{2}C - \frac{3}{2}(68B^2 + 4BC + C^2)^{\frac{1}{2}}$
	^3G	$9B + 4C$
	^1I	$6B + 6C$
d^5	^6S	0
	^4G	$10B + 5C$
	^4P	$7B + 7C$
	^4D	$17B + 5C$
	^2I	$11B + 8C$
	^4F	$22B + 7C$
	....	.....

The energy of these levels, E, can be evaluated in terms of the parameters Δ, B, and C. Using parameters B and C, which do not differ from their free ion values, we may plot the energy levels as a function of the strength of the ligand field, Δ, in the so-called Orgel diagrams. See Figure 97.

For the specification of the separation of the levels only parameters B and C are required because the parameter A involves virtually only a uniform shift of all the levels. It must further be mentioned that the 10-n electrons (viz. holes) give rise to the same number of states as the n-electrons, which means that the quantum numbers for the 10-n electron states are congruent with the quantum numbers for the n-electrons. The distributions of the levels in the two cases are thus congruent too.

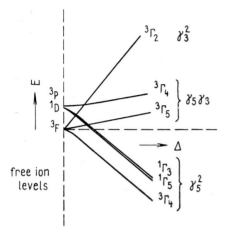

Fig. 97 Orgel diagram for the d^2-configuration in an octahedral field.

The *calculations of the energy levels of d^n-systems* are based upon two limiting conditions:

(a) in the calculations of the so-called *weak field approach* the levels of energy are considered to be defined mostly by the interactions between the electrons. The ligand field is treated as a perturbation term which splits the states of the free ion but does not cause them to interact with each other. The results of this calculation are expressed as a set of energy matrices each corresponding to a particular $^{2S+1}\Gamma_n$ state.

(b) in the calculations of the so-called *strong field approach* the levels of energy are considered to be determined particularly by the ligand field and the electrons assigned to ligand field configurations $\gamma_5^a \gamma_3^b$ (where $a + b = n$, the number of d-electrons). In the strong field approximation the electrostatic interaction is treated as a perturbation that splits the ligand field configurations but does not cause the intermixing of the different configurations. A complete calculation can be made by including the interaction between the states that have, owing to electrostatic interaction, the same **S**'s and Γn's. The results of the computations are again put in the form of a set of energy matrices, each of which corresponds to a particular $^{2S+1}\Gamma_n$. The non-diagonal elements of the matrices contain the Racah parameters, which represent the interaction between the different ligand field states $\gamma_5^a \gamma_3^b$.

137

Tanabe and Sugano [38] evaluated the complete matrices of the strong field diagonal energy for all d^n-systems in octahedral symmetry. They made use of the following notation:

$$10 \text{ Dq} \sim \Delta,$$
$$d\varepsilon \quad \sim \gamma_5,$$
$$d\gamma \quad \sim \gamma_3.$$

The matrix for the d^n-system in tetrahedral symmetry is thus the d^{10-n} matrix of octahedral symmetry, matrix d^n being applicable for d^{10-n} if the sign of Dq is changed.

The energy matrices for the d^n-systems with B, C, and C/B values are quoted in Tanabe and Sugano [38], Bates [2], and Wong and Angell [34]. For the complexes of a given ion the authors assumed a value of C/B, which is taken to be that of the free ion, although B and C may differ from their free ion values. They measured the energy levels with the help of E/B, taking Dq/B as a variable parameter, and presented them in the form of energy diagrams with E/B as ordinate and Dq/B as abscissa obtained by the solution of the above quoted determinants for the lower levels of importance. (Figures (98–104).

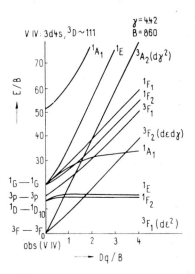

Fig. 98 Energy diagram for the d^2-configuration after Tanabe and Sugano.

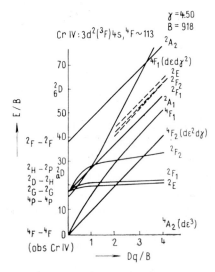

Fig. 99 Energy diagram for the d^3-configuration after Tanabe and Sugano.

The dashed lines in the range $Dq/B > 2$ are other interesting levels, which have only been solved for the strong fields approximation. The value of $\gamma = C/B$ used in the calculation is given in the upper right corner of the diagrams, together with the B value for the free ion. The levels of the free ion are given on the left-hand side of the diagrams (for comparison with theoretical levels). The energy of the lowest

138

level next to the higher configuration $3d^{n-1}\,4s$ of the free ion can be found quoted in units of B in the upper left corner of the diagrams. The ground state of the electrons in the atom is always marked on the Dq/B abscissa, according to Hund's rule, viz.

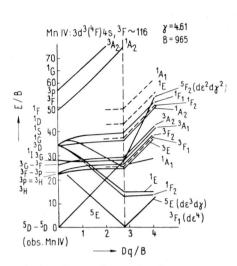

Fig. 100 Energy diagram for the d^4-configuration after Tanabe and Sugano.

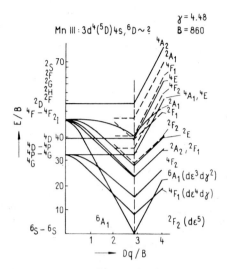

Fig. 101 Energy diagram for the d^5-configuration after Tanabe and Sugano.

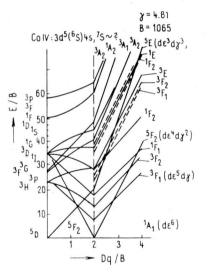

Fig. 102 Energy diagram for the d^6-configuration after Tanabe and Sugano.

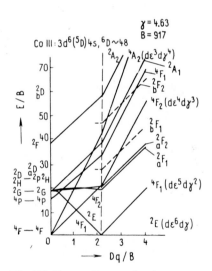

Fig. 103 Energy diagram for the d^7-configuration after Tanabe and Sugano.

the maximum number of unpaired electrons spinning parallelly. Owing to exchange interactions, this state is the most stable. In the d^4-, d^5-, d^6-, and d^7-systems, however, with sufficiently high Dq/B values, the stability of the electrons occupying the

139

low-energy orbitals γ_5 at the expense of the pairing spins is enhanced and the ground state will be the one with a lower S value. As the result of an increase of the ligand field, the ground state changes from a high to a low spin, so that the Pauling magnetic criterion distinguishing "ionic" and "covalent" complexes in terms of their magnetic susceptibility is invalidated.

In the diagrams this change in the ground state is indicated by a vertical line corresponding to the respective value of Dq/B. The intercepts of the levels with a vertical between the ground states and the excited states give the energies of the transitions and, consequently, the frequencies of the absorption bands.

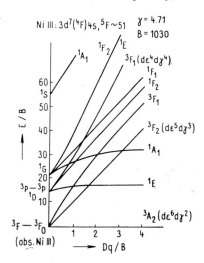

Fig. 104 Energy diagram for the d^8-configuration after Tanabe and Sugano.

As already mentioned above, transitions $d^n - d^n$ are forbidden by Laporte's rule. In the absence of a centre of symmetry in the complex the $3d$-orbitals intermingle with the higher $4p$-orbitals. Since the $d \rightarrow p$ transition is permitted, normally forbidden transitions may then occur, their intensity being of course lower.

In octahedral complexes manifesting a centre of symmetry the mixing of the d- and p-orbitals is brought about by *molecular vibrations* if the central ion is not centrally positioned. These transitions do not exhibit high intensity either.

In tetrahedral complexes that do not possess a centre of symmetry the intensity of the transitions is much higher, but still weak if compared with the Laporte permitted transitions. The relatively higher intensity of these transitions is taken to be the effect of the mixing of the $3d$-orbitals with the ligand orbitals as well as with the $4p$-orbitals.

The energy diagrams also take account of the fact that some bands in the spectra of the complexes of the transition metals are broad, other narrow. For the values of Dq/B greater than 2 the slopes of the levels given in the energy diagrams differ by integers, i.e., 0, 1, 2, etc. corresponding to the different ligand field states $d\varepsilon^a d\gamma^b$. For the levels that are not parallel to the ground level the transition energy is highly

140

dependent upon Dq/B, whereas for the excited levels (running parallel with the ground level) the transition energy is almost independent of Dq/B.

In conformity with the Frank-Condon principle, transition to various vibrational levels of the excited states will take place when the separation of the ligand-central ion changes due to thermal vibration. In this case Dq will vary about the mean value and will give rise to broad bands (if the levels are not parallel) and, vice versa, it will produce an almost line absorption if the levels are parallel.

Other aspects of the complexes of the transition metals are dealt with in Bates [2], where the reader can find the most important interpretations concerning the subject, e.g., *Tsuchida's spectrochemical series* describing the arrangement of the ligands in terms of their spectra displacing effect. The strength of the ligand field, Δ, increases from left to right:

$$I^- < Br^- < Cl^- < OH^- < NH_3^- < F^- < \text{oxalate} < H_2O < SCN <$$
$$< NH_3 < \text{ethylendiamine} < NO_2 \lll CN^-.$$

The rule of the *"average environment"* further states that if the Δ-values of the ligands differ a great deal, a substantial field component can manifest lower symmetry, in which case the absorption bands known from the regular octahedral complexes may split into a number of components.

The atoms outside the first coordination sphere of the ligands can shift the $d^n \rightarrow$ $\rightarrow d^n$ absorption bands of a given complex by $1-2\%$ only.

The *Jahn-Teller effect* postulates that only complexes of ions of the transition metals with non-degenerate ground states can manifest regular octahedral symmetry, whereas all other systems must undergo distortion from the regular octahedral symmetry to rhombic, or tetragonal, symmetry. This leads to a splitting of the ground state into a new non-degenerate ground state of lower energy.

The complexes of the transition metals are stabilized by the ligand field owing to their preference for the occupation in the ground state of the low energy γ_5^a orbitals for octahedral symmetry and γ_3^b orbitals for tetrahedral symmetry. In a high spin complex the ground state of which corresponds to the ligand field configuration $\gamma_5^a\gamma_3^b$, this stabilization energy for octahedral symmetry equals $(.4a-.6b)\Delta$ and for tetrahedral symmetry $(.-4a+.6b)\Delta^1$. For the d^2 and d^7 octahedral systems and the d^3 and d^8 tetrahedral systems this holds only for $\Delta \geqslant 20B$, viz. in the strong field approximation; but for actual complexes this is a reasonably good approximation. A number of other factors affect the stabilization of the ligand fields, e.g., the radius of the ion of the transition metal, and the radius and polarizability of the ligands. But these problems are discussed in more detail in references [2] and [34].

Molecular orbital theory

The molecular orbital theory views the central ion and the ligands as one system characterized by a set of molecular orbitals (in much the same way that an atom is characterized by s, p, d orbitals). The molecular orbitals of a transition metal complex are derived from linear combinations of the transition metal ion nd, $(n - 1)s$, and $(n - 1)p$ orbitals, viz. $3d$, $4s$, and $4p$ in the first transition series, and the ligand outer s, p orbitals. Each nearest ligand atom is taken to have one of its three p-orbitals directed along the bond towards the central ion. This orbital and the s-orbital are denoted as σ-orbitals. The other two p-orbitals, normal to the direction of the bond, are π-orbitals.

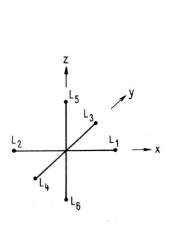

Fig. 105 Numbering of the ligands for an octahedral complex after Tanabe and Sugano.

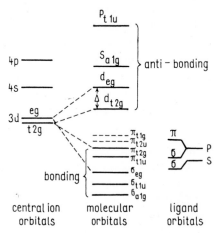

Fig. 106 Molecular orbital energy levels for an octahedral complex.

In accordance with group theory we can thus state that only the central ion and the ligand orbitals of the same symmetry can combine to establish molecular orbitals. The combination of two orbitals will produce two new orbitals, one of which is more stable (bonding orbital), the other less stable (anti-bonding orbital) than either of the original orbitals. The notation used for the ligand orbitals is as follows: σ_i denotes the σ orbital ion the i-th ligand, and $\pi_{i,r}$ the π orbital on the i-th ligand, which has its axis parallel to the X-axis. For the numbering of the ligands see Figure 105.

Figure 106 is a diagram of the energy levels of the molecular orbitals of an octahedral complex.

The orbitals of the free central ion are presented on the left; their links with the bonding and anti-bonding molecular orbitals are visualized by dashed lines for the d-orbitals. The ligand orbitals on the right are split, owing to the interactions of the electrons on the individual ligands. The order of some of the molecular orbitals may differ in different complexes from the representation shown in Figure 106; nevertheless, the principle that the diagram is based on will remain the same. The bonding and the non-bonding orbitals up to the πt_{1g} orbital are filled, the weakly anti-bonding dt_{2g} and strongly anti-bonding de_g orbitals are only partially occupied.

142

Electronic transitions may occur between the t_{2g} and e_g orbitals (cf. the description of the ligand field). But in the description of the ligand field the separation of these levels, Δ, is viewed as a measure of the perturbing effect of the ligands upon the central ion, whereas in the description of the molecular orbitals it constitutes a measure of the strongly anti-bonding character of the de_g orbitals.

The energy levels of the molecular orbitals (Figure 106) indicate that in addition to the Laporte forbidden $(g \rightarrow g)$ electron transitions between the anti-bonding dt_{2g} and de_g orbitals dealt with by the ligand field theory, there exists the possibility of the Laporte allowed $(u \rightarrow g)$ transitions involving the transfer of an electron from a predominantly ligand orbital to an orbital whose concentration is mainly on to the central ion. The spectra of these transitions, viz. the *charge transfer spectra*, manifest high intensity, which is 100 to 1000 times higher than the intensity of the ligand field spectra. Thus the lowest energy transition of this type might be from the non-bonding πt_{2u} ligand orbital to the dt_{2g} orbital for complexes with complete dt_{2g} orbitals. It is also possible that Laporte forbidden charge transfer transition may occur, for instance $\pi t_{1g} \rightarrow dt_{2g}$ or de_g, but examples of the spectra corresponding to these transitions seem to be very rare.

Although in Figure 106 the ligand π orbitals are presented as more stable than the d-orbitals and, consequently, occupied, certain ligand π orbitals are less stable than the completely filled d-orbitals. Complexes with the latter ligands can manifest charge transfer transitions in the reverse direction to those mentioned above, viz. from the dt_{2g} or de_g orbitals to the ligand π orbitals.

Two more types of transitions occur, viz. *transitions between the ligand orbitals*, resulting in a re-distribution of the electrons on the ligands, and transitions within the central ion.

Since this subject poses problems that are rather difficult to tackle, calculations of the molecular orbitals have so far been rare.

Bamford [3], Douglas [40], Bates [2], Turner [41], and others have attempted an application of the theory of the ligand fields to the analysis of the absorption spectra of the ions of the transition metals in binary and multi-component glasses. Table 26 presents a survey of the absorption bands of the ions of the transition metals, the likely coordinations, and the respective electron transitions in Bethe's notation.

The theoretical foundations of the oxidation-reduction equilibria of the transition elements in glasses associated with the change in the coloration were laid by Douglas, Nath, Paul, and Edwards [52, 53, 54, 55, 56, 57, 58]. For details, the reader is referred to these references.

Lambert-Beer's law

In propagating through glass, the radiant flux, Φ, is attenuated due to absorption. This attenuation is proportional to the path covered, l, and to the concentration of the absorbing centres, c.

Table 26

Ion	Coordination	Absorption band λ (nm)	Transition
Cu^{2+}	octahedral	780	$^2\Gamma_3(D) - ^2\Gamma_5(D)$
Ni^{2+}	octahedral	450	$^3\Gamma_2(F) - ^3\Gamma_4(P)$
		930	$^3\Gamma_2(F) - ^3\Gamma_4(F)$
		1800	$^3\Gamma_2(F) - ^3\Gamma_5(F)$
	tetrahedral	560	$^3\Gamma_4(F) - ^3\Gamma_4(P)$
		630	$^3\Gamma_4(F) - ^3\Gamma_4(P)$
		1200	$^3\Gamma_4(F) - ^3\Gamma_2(P)$
Co^{2+}	tetrahedral	530	$^4\Gamma_2(F) - ^4\Gamma_4(P)$
		590	$^4\Gamma_2(F) - ^4\Gamma_4(P)$
		645	$^4\Gamma_2(F) - ^4\Gamma_4(P)$
Fe^{2+}	octahedral	1050	$^5\Gamma_5 \quad - ^5\Gamma_3$
Fe^{3+}	tetrahedral	435	
		420	
		380	
Mn^{3+}	octahedral	490	$^5\Gamma_3(D) - ^5\Gamma_5(D)$
Cr^{3+}	octahedral	450	$^4\Gamma_2(F) - ^4\Gamma_4(F)$
		650	$^4\Gamma_2(F) - ^4\Gamma_5(F)$

For the relationship between the incident and the unabsorbed and emerging radiant flux we can write, after Lambert,

$$\ln\left(\frac{\Phi_{tr}}{\Phi}\right)_\lambda = -al,$$ (178)

where a is the linear absorption coefficient,

l the thickness of the glass,

Φ the incident radiant flux,

and Φ_{tr} the unabsorbed radiant flux emerging from the glass.

Lambert's law can then be re-written in the form

$$\tau_\lambda = \left(\frac{\Phi_{tr}}{\Phi}\right)_\lambda = e^{-al}.$$ (179)

Introducing the common logarithm, we get

$$\log\left(\frac{\Phi_{tr}}{\Phi}\right)_\lambda = -a'l.$$ (180)

where $a' = 0.4343 \cdot a$.

Then $a'l = 0.4343al$. (181)

For the ratio $- \log\left(\dfrac{\Phi_{tr}}{\Phi}\right)_\lambda$ the terms used are: internal optical density, D_i, or extinction, E. We can thus write

$$D_i = -\log \tau_i = \log \frac{1}{\tau_i}. \tag{182}$$

Giving the Lambert equation (179) the form

$$-\log\left(\frac{\Phi_{tr}}{\Phi}\right)_\lambda = a'l = D_i, \tag{183}$$

we can write for a:

$$a = \frac{D_i}{l \cdot 0.4343}. \tag{184}$$

The numerical values of the linear absorption coefficient a is generally related to the wavelength (nm, μm] or energy, E [eV], and is made use of in the physical interpretation of the absorption spectra. Its usual quotation is in cm^{-1}.

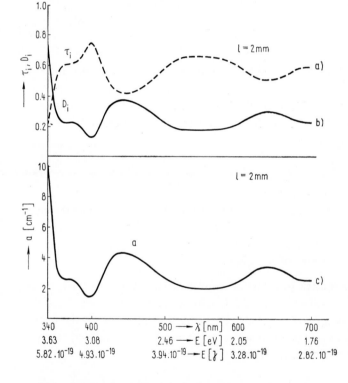

Fig. 107 Spectral transmission (a), internal transmission density (b), and linear absorption coefficient (c) of sun glass.

Figure 107 shows curves of the spectral transmission τ (a), internal optical density, D_i (b), and the linear absorption coefficient a, (c) related to the wavelength for sun glass. The values of energy in joules [J] and in electronvolts [eV] are quoted for reference.

145

Beer derived a similar law for the ratio of the unabsorbed to the incident radiation flux related to the concentration of the absorption centres, c:

$$\log\left(\frac{\Phi_{tr}}{\Phi}\right)_\lambda = -a'c, \tag{185}$$

where c stands for the concentration of the absorption centres.

Lambert-Beer's law is thus a combination of the two cited laws, viz.

$$\log\left(\frac{\Phi_{tr}}{\Phi}\right)_\lambda = -a'cl. \tag{186}$$

Measuring spectral transmittance in glasses, we obtain the values of the so-called transmission factor:

$$\tau = \left(\frac{\Phi_{tr}}{\Phi}\right)_\lambda. \tag{187}$$

In practical computations use is also made of the expression

$$a' = \frac{D_i}{cl} \tag{188}$$

giving the linear absorption coefficient by expressing numerically the internal optical density for the constant concentration of the absorbing component and constant thickness of a glass. For practical purposes Lambert's relationship is currently made use of in the computations of spectral transmittance for various thicknesses in the following form:

$$\log\frac{\Phi_2}{\Phi} = \frac{l_2}{l_1}\log\frac{\Phi_1}{\Phi}, \tag{189}$$

where Φ_1, Φ_2 denote the radiant flux emerging from the glass of thicknesses l_1 and l_2,

Φ is the incident radiant flux,

and l_1, l_2 denote the thicknesses of the glass.

The graphs are usually plotted in terms of transmission factor τ, internal transmittance τ_i, absorptance α, internal optical density D_i (extinction), or the linear absorption coefficient a as a function of the wavelength in nm, µm, or the energy in joules or eV. For optical loss (optical waveguide technology) is related to $D_i =$ $= 10\,dB$ (where dB denotes the decibel).

Definition of the transmission and absorption of radiation.

We define the *absorption of radiation* as the conversion of radiant energy into another type of energy due to the interaction of the radiation with a medium (glass). Transmission is defined by transmission factor τ, which is dependent upon the wavelength, the angle of incidence, and the polarization of the radiation. It is defined by the relationship

$$\tau = \left(\frac{\Phi_{tr}}{\Phi}\right)_\lambda .$$

(190)

The *internal transmittance*, τ_i, if scattered transmission is absent, is the ratio of the emergent flux to the incident flux. The absorption factor, α, can be defined as the ratio of the flux absorbed by the medium to the incident flux. The absorption factor is also dependent upon the wavelength, the angle of incidence, and the polarization of the radiation. It is defined by the relationship

$$\alpha = \left(\frac{\Phi_\alpha}{\Phi}\right)_\lambda .$$

(191)

Internal absorptance α_i is the ratio of the radiant, or luminous, flux absorbed between the entry and the exit surfaces of the medium to the incident flux.

Optical density D is defined as the common logarithm of the reciprocal value of the transmission factor, viz.

$$D = -\log \tau,$$

(192)

and *internal optical density* D_i as the common logarithm of the reciprocal of internal transmittance, viz.

$$D_i = -\log \tau_i .$$

(193)

The *linear absorption coefficient*, a, for monochromatic radiation is the ratio of internal absorptance for an infinitely thin layer of the material to that for a layer of thickness l. It can be written from Lambert's law as

$$a = \frac{D_i}{l \cdot 0.4343} .$$

(194)

4.3.3   Absorption of radiation in glasses

From the curves of spectral transmission presented in Figure 108 it follows that in the amorphous materials in the vitreous state the spectral transmission of radiation in the short wave region of the spectrum is approximately limited by the wavelength $\lambda = 160$ nm. The measurement of transmission in high purity quartz glass must be undertaken in an atmosphere of dry nitrogen. In this case spectral transmission in the infra-red region of the spectrum is limited by the wavelength of approximately $\lambda = 5$ nm.

However for window glass of composition $Na_2O - K_2O - CaO - MgO - SiO_2$ with absorption in the infra-red region of the spectrum ($\lambda = 5$ µm) approximately uniform, the absorption edge in the ultraviolet region shifts into the visible region, towards $\lambda = 300$ nm.

Chalcogenide glasses, on the other hand, are opaque to visible radiation, because the absorption edge has shifted into the infra-red region, towards $\lambda = 900$ nm. But they transmit infra-red radiation up to about $\lambda = 20\ \mu m$, according to the particular composition.

Absorption of radiation in the ultra-violet region of the spectrum

The absorption of radiant energy in the ultra-violet region of the spectrum is caused by the resonance of the radiation with the frequencies of the electrons, as stated in the theoretical part of the present book. If the material (e.g., a metal) contains free electrons, the resonance will reach its peak and the material will be non-transmissive for the radiation of the optical region.

The resonance need not however, be the result of the interaction of the radiation with the free electrons only: radiation can also interact with the other electrons of various levels of bonding. The location of the absorption edge in the ultra-violet region of the spectrum thus depends upon the composition of the glass, impurities, structural defects, colour centres, etc.

In relatively pure quartz glass the absorption edge is placed approximately at the wavelength of $\lambda = 160$ nm (~ 8 eV) in the ultra-violet region of the spectrum (see Figure 109).

Table 27

Glass	Reflection maximum [eV]			
SiO_2 (crystal)	—	—	10.2	11.5
SiO_2 (glass)	—	—	10.2	11.5
$Li_2O . 2 SiO_2$	8.5	9.3	—	11.5
$Na_2O . 2 SiO_2$	8.5	9.3	—	11.5
$Na_2O . 3 SiO_2$	8.5	9.3	—	11.5
$Na_2O . 6 SiO_2$	—	9.0	—	11.5
$Na_2O . CaO . 5 SiO_2$	8.5	9.3	—	11.5
$Na_2O . Al_2O_3 . 3 SiO_2$	8.5	—	10.2	11.5

This kind of relatively pure quartz glass (curve I) contains the following impurities (in p.p.m.):

K < 0.1   Al < 0.1   Cr < 0.1   Co < 0.1
Li < 0.1   Fe < 0.1   Ni < 0.1   Zn < 0.1

　　Ca < 0.1   Ti < 0.1   Cu < 0,1
　　Mg < 0.1   Na < 0.2   Mn < 0.1

When measuring the reflectivity of high purity quartz glass Philipp [42] found absorption maxima at 10.2 eV, 11.7 eV, 14.3 eV, and 17.2 eV, as shown in Fig. 110.

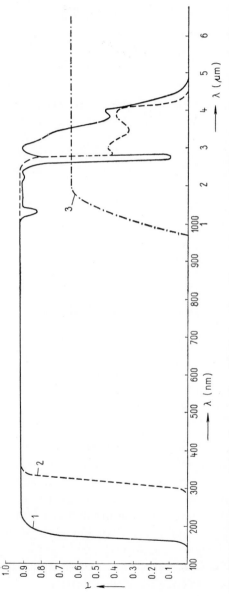

Fig. 108 Spectral transmission of quartz glass (*1*) window glass (*2*), and Ge₃₄As₈Se₅₈ chalcogenide glass (*3*) in the optical region of the spectrum.

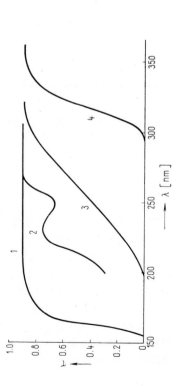

Fig. 109 Spectral transmission of relatively pure quartz glass (*1*), impure quartz glass (*2*), the $Na_2O \cdot 3SiO_2$ glass melted from extremely pure raw materials (*3*), and window glass (*4*) in the ultra-violet region of the spectrum.

149

Table 27 (after Siegel [43]) lists the reflective absorption maxima of high purity quartz crystal, of synthetic quartz glass produced by Corning (No. 7940), and of various high purity binary and ternary silicate glasses.

It can safely be assumed that the differences between the locations of the absorption maxima quoted by different authors are due to the different purity of the glasses and the quartz crystal. Curve 2 in Figure 109 is the spectral transmission in quartz glass containing (in p.p.m.):

Al = 50	Li = 2	Cu < 0.1	Fe = 0.5
Ca = 2	Mn < 0.1	Ti < 0.8	Mg = 0.1
Co < 0.1	Na = 1	Cr < 0.1	Zn < 0.1
K = 0.2	Ni < 0.1.		

The introduction of impurities will thus shift the absorption edge towards the near visible region of the spectrum.

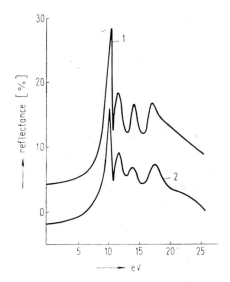

Fig. 110 Reflection spectra of β-quartz (1) and of quartz glass (2).

The introduction of further high concentration components into silicon oxide will destroy the relatively firm oxygen-silicon bridges, new non-bridging bonds will be established, mostly of an ionic character, and electrons will be more weakly bonded. We can therefore expect the absorption edge to move towards the visible region of the spectrum. Curve 3 in Figure 109 represents glass of the $Na_2O . 3SiO_2$ system melted from high purity raw materials, and curve 4 window glass of the following composition (weight per cent): $SiO_2 = 71.9$, $Al_2O_3 = 0.9$, $CaO = 7.3$, $MgO = 3.9$, $NaO + K_2O = 15.3$, $Fe_2O_3 = 0.06$. Here, the location of the absorption edge is in the close vicinity of the visible region of the spectrum.

According to Siegel [43, 61], we can assume that the absorptivity of quartz glasses in the ultra-violet region of the spectrum is caused

(a) in the 350−200 nm waveband, by the impurities present in the glass;

(b) in the 200−150 nm waveband, by induced absorption due to the disruption of the relatively firm Si − O bridges;

(c) in the range of the wavelengths shorter than 150 nm, by the relatively firm Si − O bridges of the basic structure of quartz glass.

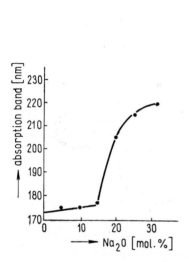

Fig. 111 Shift of the absorption edge in the Na_2O—B_2O_3 glass as a function of Na_2O content.

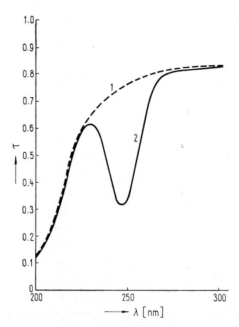

Fig. 112 Typical absorption band in quartz glass characterizing non-stoichiometry of SiO_{2-x} (2) as compared with stoichiometric SiO_2 glass (1).

Vitreous boron trioxide with an absorption edge at 170 nm transmits well in the ultra-violet region of the spectrum. According to McSwain and Borelli [60], the cause of this ultra-violet transmission are the $[BO_3]$ and $[BO_4]$ coordinations with firmly bonded oxygen. Introducing alkalis will produce non-bridging oxygens, and the absorption edge with 30 mol. % Na_2O will approach $\lambda = 220$ nm (see Figure 111).

These problems are dealt with in greater detail from the point of view of the structure in Wong and Angell [34], Ruffa [44], Philippe [42], Siegel [61, 43], Krause [45] and elsewhere.

The real structure of quartz glass reduced by melting in a graphite crucible differs from the theoretically postulated structure in manifesting non-stoichiometric silicon coordinations, viz. a lower degree of oxidation. According to Garino-Canina [5], who assume the reaction

$$SiO_2 + C \rightarrow SiO + CO, \tag{195}$$

the gaseous silicon monoxide and carbon monoxide produced at melting temperatures above 1327° C can reduce silicon dioxide to SiO, Si, or to Si_2O_3, Si_3O_4 (mixtures of SiO_2 + SiO + Si). This reduced non-stoichiometric quartz glass corresponds, according to [5], to formula SiO_{2-x}, which has a characteristic absorption band at $\lambda = 242$ nm (see Figure 112).

Absorption of radiation in the visible region of the spectrum

Glasses that do not contain absorbing components are transmissive for radiation in the visible region of the spectrum; of course, the transmitting capacity depends upon the degree of the purity of the raw materials used. The reduction in spectral transmittance during measurement is caused by the reflection of the radiation on the two functional surfaces (see Section 4.1), so that subtracting the losses due to reflection, we should obtain the theoretical values of spectral transmittance.

Spectral transmission in the visible region is reduced, either uniformly or selectively, if impurities are present in the glass, if colouring components are introduced, if the glass is exposed to irradiation, etc. This fact is generally taken advantage of if coloured glasses are to be produced that will absorb certain wavelengths of the visible region of the spectrum. In glasses, absorption is generally seen as broad bands. However, for the rare earths, where the electrons undergo transitions within inner orbits, we get relatively narrow absorption bands.

According to the size of the so-called colour centres we can classify the colourants (after [1]) into the following groups:

Group A: The colour centre is present in the form of simple or more complex ions, or molecules, up to about 1 nm in size. The group comprises the following elements: nickel, cobalt, iron, manganese, chromium, copper, vanadium, titanium, neodymium, cerium, praseodymium. And the dependence of absorption upon the concentration of the colour centres fully satisfies Lambert-Beer law. For further information on the absorption of the ions of the transition metals the reader is referred to Section 4.3.2 dealing with the ligand field theory.

Group B: The colour centre is present in the form of colloidal particles up to about 1 − 500 nm in size. They are mostly particles of silver, gold, copper, etc., and Cd (S, Se) compounds. As the particles grow during thermal treatment, the glass changes its colour. A higher concentration and a larger size of the particles will give rise to scatter.

Group C: The colour centres are produced by microscopic, or larger, particles suspended in the glass, crystallized and separated from the glass during cooling or thermal treatment. Such glasses manifest cloudiness and they scatter light.

Group D: The colour centres are produced by radiation, ultra-violet radiation,

152

γ-radiation, β-radiation, X-ray radiation, etc. For further information the reader is referred to Section 6.4, where the problem is taken up in detail.

A very interesting way of dealing with the problem of colour centres is Weyl and Marboe's [8] approach. These authors were the first to emphasize the necessity of taking into account the effect of the surrounding fields of force in the theoretical interpretation of the absorption bands. We shall now mention briefly some of their views concerning the colouration of glasses.

According to Weyl and Marboe [8] the colouration of inorganic materials and of glasses is caused by so-called *inorganic chromophores*. It is a well-known fact (cf. Section 4.3.2) that one of the most important factors is the distribution of the electrons; this distribution is in turn affected by the fields of force of the surrounding ligands. The colour of inorganic materials and of glasses is also affected by the type of defects in the lattice or in the structural arrangement.

Weyl and Marboe [8] have classified inorganic chromophores into the following groups:

A. Atoms and molecules

These chromophores do not possess a charge and are therefore less sensitive to the surrounding fields of force. This is manifested by the fact that the colour shade is almost independent of the composition of the glass. For instance, atomically dispersed silver or gold do not cause visible colouration of the glass. However, the association of such atoms into metallic particles gives rise to colour centres (see H. below).

For instance, S_2 molecules in alkaline phosphate or borate glasses give these glasses a bluish tint; the colour centre is thus not dependent upon the surrounding fields of force to any major degree.

On the other hand, if the S_2 centre undergoes a transition to polysulphides $(S_n)^{2-}$, dark borwn shades will appear. Colour generation caused by selenium and tellurium is of the same type.

B. Ions of the transition elements

The problems of the colour centres made up of ions of the transition metals have been dealt with in detail in Section 4.3.2.

C. Mutual deformation of coulourless ions

The colourless ions, too, e.g., Cd^{2+}, Pb^{2+}, can generate colour in glasses, provided the most polarizable anions, e.g., S^{2-}, Se^{2-}, Te^{2-}, are also present. The mutual interaction and deformation of the electron orbitals, e.g., of the sulpho-selenides of cadmium, give rise to red colours. The latter is to a high degree dependent upon the ratio of the components, viz. upon the mutual deformation of the electron orbitals.

153

D. Interaction of valencies

Typical examples of these colour centres are the polyvalent ions present in glasses, e. g., Fe^{2+}, Fe^{3+}, Cr^{6+}, Cr^{3+}, Mn^{2+}, Mn^{3+}, Cu^+, Cu^{2+} etc. For instance, an ion of a valency that does not give rise to a colour centre in the visible region can influence the colouring capacity of the ions present in another valecny (e.g., $Fe^{2+} \rightleftarrows Fe^{3+}$). Of course, in these cases the resulting colour will also be affected by the field of force of the surrounding medium.

E. Induced valency

The effect of the induced valency is based upon the fact that an ion present in minute quantities may take the place of an ion contained in the structure and is thus also forced to accept the respective valency. It will develop a colour different from the one it possessed in its original state.

F. Symmetric and asymmetric formations

The structure of the basic glass forming oxide SiO_2 is symmetric owing to the relatively very powerful energy fields of the central ion Si^{4+}, though the O^{2-} anion may otherwise be less stable. On the other hand, symmetry of such a high order is not exhibited by, for example, the surface of the SiO_2 crystal. Thus the $[SiO_4]^{4-}$ tetrahedron can undergo high deformation, producing a colour centre of the $Si^{4+} \left[\dfrac{O^{2-}}{2} \right]_3 O^-$ type.

This theory also provides us with the explanation of why, in the case of the Pb^{2+} ions in heavy lead glasses, colouration appears when these glasses are subjected to thermal treatment. This is a case of the fluctuation of symmetry.

A similar effect of asymmetry on the colour generation in glasses can be observed in the case of phase-separated glasses.

Fig. 113 Spectral transmission of glasses containing ionic and molecular colourant:
(a) — $6 SiO_2 . CaO . Na_2O$ and $6 SiO_2 . CaO. K_2O$ glasses (*1* and *2*, respectively) coloured with 0.25 per cent of NiO ($l = 2$ mm); (b) — $6 SiO_2 . CaO . K_2O$ glass coloured with 0.05 % of CoO ($l = 2$ mm); (c) — $6 SiO_2 . CaO . Na_2O$ glass coloured with 2 % of Fe_2O_3 (*1*) and 2 % of FeO (*2*) ($l = 2$ mm); (d) — $6 SiO_2 . CaO . Na_2O$ glass coloured with 2 % of $KMnO_4$ ($l = 2$ mm); (e) — $6 SiO_2 . CaO . Na_2O$ glass coloured with 1 % of $K_2Cr_2O_7$ ($l = 2$ mm); (f) — $6 SiO_2 .$. CaO . K_2O glass coloured with 1 % of $Na_2U_2O_7$ ($l = 2$ mm); (g) — spectral transmission in gold ruby, after 1 hour at 600 °C ($l = 2$ mm); (h) — $6 SiO_2 . CaO . Na_2O$ glass coloured with 4 % of V_2O_5 ($l = 2$ mm); (i) — $6 SiO_2 . CaO . Na_2O$ glass coloured with 1 % of CuO ($l = 2$ mm); (j) — $6 SiO_2 . PbO . Na_2O$ glass coloured with 2 % of CeO_2 ($l = 2$ mm); (k) — $6 SiO_2 . CaO .$. Na_2O glass coloured with 10 % of Nd_2O_3 ($l = 2$ mm); (l) — $6 SiO_2 . CaO . Na_2O$ glass coloured with 10 % of Pr_2O_3 ($l = 2$ mm); (m) — alkali-zinc-silica glass coloured with 1.5 % of CdS, 0.15 % of Se, and 0.15 % of Na_2SeO_3 (*1*), alkali-zinc-silica glass coloured with 1 % of CdS, 0.3 % of Se, and 0.8 % of Na_2SeO_3 (*2*), and sodium-potassium-lime-silica glass coloured with 0.03 % of Se (*3*); (n) — sodium-lime-silica glass coloured with 0.2 % of $AgNO_3$ (*1*), copper ruby coloured with 1.2 % of Cu_2O (*2*) ($l \doteq 2$ mm).

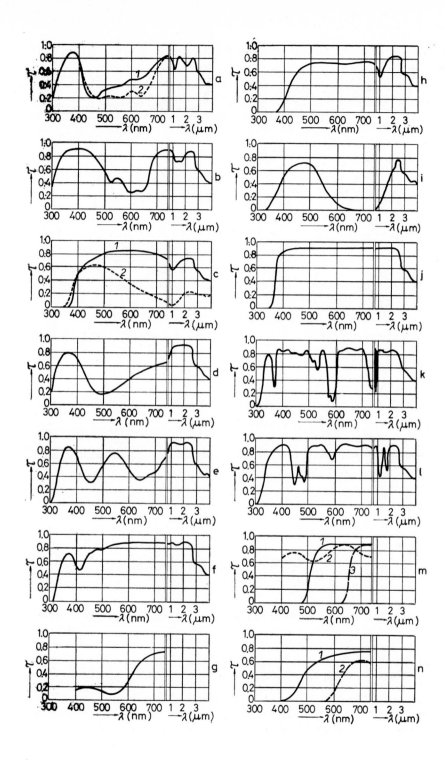

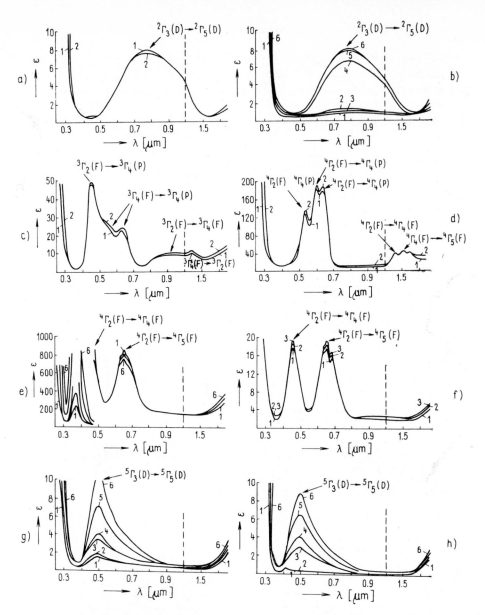

Fig. 114 Absorption spectra of ionic colourants in Na_2O—SiO_2 glass. Glasses melted under oxidizing and reducing conditions:

(a) — oxidizing conditions —

Na_2O	21.8 wt. %	CuO	0.42 wt. % (1)	SiO_2	57.8 wt. %
SiO_2	77.8 wt. %	Na_2O	41.6 wt. %	CuO	0.56 wt. % (2)

(b) — reducing conditions —

Na_2O	15.6 wt. %	CuO	0.12 wt. % (1)	SiO_2	79.3 wt. %
SiO_2	84.3 wt. %	Na_2O	20.6 wt. %	CuO	0.16 wt. % (2)

156

G. Transfer of an electron by radiation

Under this heading are grouped photochemical reactions in, for example, photo-sensitive, photoform, photochromatic, and polychromic glasses caused by solarization, the action of γ-radiation, X-ray radiation, β-radiation, and by other types of energy-intensive radiation.

This subject will be discussed in more detail in Chapter 6.

Na_2O	26.1 wt. %	SiO_2	69.3 wt. %	CuO	0.24 wt. % (5)
SiO_2	73.8 wt. %	CuO	0.15 wt. % (4)	Na_2O	39.6 wt. %
CuO	0.25 wt. % (3)	Na_2O	35.8 wt. %	SiO_2	60.3 wt. %
Na_2O	30.6 wt. %	SiO_2	64.1 wt. %	CuO	0.14 wt. % (6)

(c) — oxidizing conditions —

Na_2O	16.6 wt. %
SiO_2	83.0 wt. %
NiO	0.19 wt. % (1)

(d) — oxidizing conditions —

Na_2O	20.9 wt. %	Na_2O	39.4 wt. %
SiO_2	79.0 wt. %	SiO_2	60.0 wt. %
Co_3O_4	0.054 wt. % (1)	Co_3O_4	0.070 wt. % (2)

(e) — oxidizing conditions —

Na_2O	16.4 wt. %	Na_2O	26.5 wt. %	Na_2O	36.1 wt. %
SiO_2	83.5 wt. %	SiO_2	73.5 wt. %	SiO_2	63.8 wt. %
Cr_2O_3	0.09 wt. % (1)	Cr_2O_3	0.08 wt. % (3)	Cr_2O_3	0.05 wt. % (5)
Na_2O	20.7 wt. %	Na_2O	31.0 wt. %	Na_2O	39.6 wt. %
SiO_2	79.0 wt. %	SiO_2	69.0 wt. %	SiO_2	60.4 wt. %
Cr_2O_3	0.08 wt. % (2)	Cr_2O_3	0.1 wt. % (4)	Cr_2O_3	0.03 wt. % (6)

(f) — reducing conditions —

Na_2O	15.2 wt. %	Na_2O	20.7 wt. %	Na_2O	25.5 wt. %
SiO_2	84.7 wt. %	SiO_2	79.2 wt. %	SiO_2	74.4 wt. %
Cr_2O_3	0.11 wt. % (1)	Cr_2O_3	0.10 wt. % (2)	Cr_2O_3	0.11 wt. % (3)

(g) — oxidizing conditions —

Na_2O	17.2 wt. %	Na_2O	26.5 wt. %	Na_2O	35.0 wt. %
SiO_2	82.1 wt. %	SiO_2	71.8 wt. %	SiO_2	63.1 wt. %
Mn_3O_4	0,88 wt. % (1)	Mn_3O_4	1.20 wt. % (3)	Mn_3O_4	1.30 wt. % (5)
Na_2O	22.2 wt. %	Na_2O	30.6 wt. %	Na_2O	40.8 wt. %
SiO_2	76.9 wt. %	SiO_2	67.5 wt. %	SiO_2	58.2 wt. %
Mn_3O_4	0.89 wt. % (2)	Mn_3O_4	1.20 wt. % (4)	Mn_3O_4	1.30 wt. % (6)

(h) — reducing conditions —

Na_2O	15.7 wt. %	Na_2O	29.7 wt. %	Na_2O	25.3 wt. %
SiO_2	73.5 wt. %	SiO_2	69.3 wt. %	SiO_2	73.9 wt. %
Mn_3O_4	0.83 wt. % (1)	Mn_3O_4	1.04 wt. % (4)	Mn_3O_4	0.80 wt. % (3)
Na_2O	20.7 wt. %	Na_2O	35.0 wt. %	Na_2O	39.1 wt. %
SiO_2	78.3 wt. %	SiO_2	64.0 wt. %	SiO_2	60.0 wt. %
Mn_3O_4	0.98 wt. % (2)	Mn_3O_4	1.16 wt. % (5)	Mn_3O_4	1.07 wt. % (6)

H. Metals in glass

Metals, e.g., silver and gold, present in molten glass that is subjected to fast cooling take the form of electro-neutral atoms, $Ag°$, $Au°$, which do not generate colour in glass but are the cause of its typical scattering capacity.

Owing to their electro-neutrality they are not influenced to any major degree by the surrounding field of force. At elevated temperatures they can move diffusely and cluster into larger colloidal agglomerates of the $(Ag°, Au°)$ type and hence give colour.

A partial effect of the outer fields is the fact that, if these forces are weak, striking sets in at lower temperatures, and vice versa.

The above classification and analysis of colour in glasses can hardly cover the great variety of all possible cases; neither should the fact be disregarded that in a number of cases these problems have not been unequivocally clarified.

The problems of colour generation in glasses have been dealt with in detail by Weyl and Marboe [8], Weyl [6], Bamford [50], Vargin [46], and Kocík, Nebřenský and Fanderlik [1]. The reader is referred to the cited literature.

Figure 113 presents graphs of spectral transmission of several ionic and molecular colourants in a three-component glass $(Na_2O-CaO-SiO_2)$ after [1]. Bamford [3] presents and interprets the absorption spectra of the transition metals in two-component glasses (Na_2O-SiO_2), as shown in Figure 114.

In view of the fact that the materials applied as glass colourants differ in their dyeing capacity, Kříž [33] determined their equivalent concentrations using the sensitivity of the human eye as the criterion. He made use of a glass of the following composition (weight per cent): $SiO_2 = 74.3$, $Na_2O = 14.1$, $K_2O = 4.1$, $CaO = 4.5$, and $MgO = 3.0$. (See Table 28).

Absorption of radiation in the infra-red region of the spectrum

In the infra-red region of the spectrum the propagation of radiation through glass is limited to approximately the $\lambda \doteq 5$ μm wavelength (see Figure 108). In quartz glass it is the shortwave edge of an absorption band with maximum at $9-9.5$ μm. This absorption band is due to resonance with the fundamental frequencies of the $Si-O$ vibrations (asymmetric vibrations $\leftarrow Si-O \rightarrow \leftarrow Si$). Symmetric vibrations ($\leftarrow Si-O-Si \rightarrow$, or $Si \rightarrow O \leftarrow Si$) correspond to the absorption band at 12.5 μm, those of the $Si-O-Si$ and $O-Si-O$ type to the absorption band at 21 μm. Figure 115 indicates the vibrations considered and the graphs of spectral transmittance of quartz glass, tridymite, and α-cristobalite in the infra-red region of the spectrum (after Neurot [47]).

The curves of spectral transmittance of vitreous boron trioxide and of crystalline boric acid and B_2O_3 are presented in Figure 116 (after [47]), those of vitreous germanium dioxide and crystalline GeO_2 (in hexagonal and tetragonal forms) in Figure 117.

158

Table 28

Colourant	Concentration [wt. %] for $D = 0.1$	Colour shade	Source of colourant
CoO	0.0018	French blue	CoO
NiO	0.0078	neutral grey	NiO
Se	0.0250	pink	Na_2SeO_3
Cr_2O_3	0.0324	serpentine green	$K_2Cr_2O_7$
CuO	0.116	blue	$CuSO_4 . 5 H_2O$
MnO	0.207	violet	MnO_2
U_3O_8	0.381	pyrethrum yellow with green fluorescence	$Na_2U_2O_7$
Fe_2O_3	0.384	greyish green	Fe_2O_3
V_2O_5	0.655	serpentine green	NH_4VO_3
Nd_2O_3	0.92	purple with red fluorescence	Nd_2O_3 (90 %)
Pr_2O_3	2.38	greyish green with yellow fluorescence	Pr_2O_3 (50 %)
$TiO_2 + CeO_2$	6.6	yellow	$TiO_2 + CeO_2$ (3 : 1)

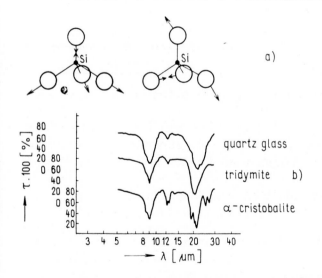

Fig. 115 Vibrations in $[SiO_4]^{4-}$ (a), spectral transmission in quartz glass, tridymite and α-cristobalite (b) in the infra-red region of the spectrum.

These characteristic absorption bands have, of course, been obtained from measurement on powdered materials mixed with potassium bromide (see Section 4.3.6).

For the usual thicknesses of glasses it is assumed that spectral transmission is limited by the absorption edge at about 5 μm.

Preceding the absorption band corresponding to the resonance of the radiation frequency with the vibrations of the Si−O bonds, three characteristic absorption bands can be observed in quartz glasses: at $\lambda = 2.73 - 2.85$ μm, $\lambda = 3.5$ μm, and $\lambda = 4.3$ μm (in synthetic fused silica at $\lambda = 1.38$ μm, $\lambda = 2.22$ μm, $\lambda = 2.73$ μm $\lambda = 3.81$ μm and $\lambda = 4.45$ μm [61]).

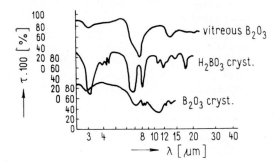

Fig. 116 Spectral transmission in vitreous B_2O_3, cryst. H_3BO_3, and B_2O_3 in the infra-red region of the spectrum.

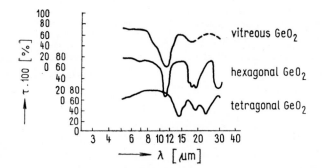

Fig. 117 Spectral transmission of vitreous GeO_2 and cryst. GeO_2 (hexagonal and tetragonal) in the infra-red region of the spectrum.

The location of each absorption band corresponds to OH groups in the structure of the glass [9, 10, 12] (see Figure 118).

The concentration of OH groups in a glass, and thus also the intensity of the absorption bands, depend upon the technology of melting, with the quantitative ratio of the individual structural arrangements determined by the composition of the glass.

The solubility of H_2O in the form of OH groups in molten glass depends upon the activity of the oxygen ions O^{2-}. For the equilibrium we can thus write

$$k_1 = \frac{a_{OH^-}^2}{a_{O^{2-}} \cdot p_{H_2O}},$$ (196)

where a denotes the activity of the ions
and p_{H_2O} the partial pressure of water vapour.

For the activity of the OH groups we can write

$$a_{OH^-} = k_1 a_{O^{2-}} \sqrt{p_{H_2O}}.$$ (197)

160

The activity of OH groups in the glass melt is thus proportional to the square root of the partial pressure over the melt. For some special purposes it is therefore essential that glasses should be melted in vacuo, in dry atmosphere, the melt flushed with dry gases, etc. Figure 119 shows the spectral transmittance of $Na_2O-K_2O-ZnO-SiO_2$ glass melted under various technological conditions.

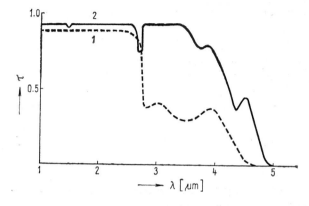

Fig. 118 Spectral transmission in window glass (1) and quartz glass (2) in the infra-red region of the spectrum with characteristic absorption bands of the OH groups.

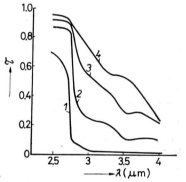

Fig. 119 Spectral transmission in $K_2O-Na_2O-SiO_2$ glasses melted under various technological conditions: 1 — melted from batch by standard technology; 2 — melted from batch sintered at 600 °C; 3 — melted from batch sintered at 600 °C, 3 % of fluorine in batch; 4 — melted from batch sintered at 600 °C, 4 % of fluorine in batch, the melt blown through with dry air during melting.

The concentration of OH groups (in p.p.m.) in quartz glasses can be computed, according to Yovanovitch [11], from the intensity of the absorption band at $\lambda = 2.73 - 2.85$ μm. If, according to Lambert-Beer's law,

$$\Phi = \Phi_0 e^{-c\varepsilon l}, \tag{198}$$

where c denotes the concentration [mol. 1^{-1}],

ε the molar linear coefficient of absorption — 77.5 [1 . mol^{-1} . cm^{-1}] and l the thickness of the glass [cm],

then

$$\log \Phi = \log \Phi_0 - (c\varepsilon l), \tag{199}$$

and

$$-\log \frac{\Phi}{\Phi_0} = c\varepsilon l = \log \frac{\Phi_0}{\Phi}. \tag{200}$$

161

For **c** we can write

$$\mathbf{c} = \frac{1}{\varepsilon l} \log \frac{\Phi_0}{\Phi} = \frac{1}{77.5} \log \frac{\Phi_0}{\Phi} \qquad [\text{mol} \cdot \text{l}^{-1}]. \qquad (201)$$

For $M \doteq 17$

$$\varepsilon = \frac{77\,500 \cdot 2.2}{17} \qquad [\text{cm}^{-1}], \qquad (202)$$

we can then write

$$c^x = \frac{17}{77\,500 \cdot 2{,}2} \cdot \frac{1}{l} \log \frac{\Phi_0}{\Phi} \qquad [\text{g} \cdot \text{g}^{-1}], \qquad (203)$$

$$c_1 = \frac{17\,000}{77.5 \cdot 2.2} \cdot \frac{1}{l} \log \frac{\Phi_0}{\Phi} \qquad [\text{p.p.m.}], \qquad (204)$$

$$c_1 \doteq \frac{100}{l} \log \frac{\Phi_0}{\Phi} \qquad [\text{p.p.m.}], \qquad (205)$$

where c_1 is the concentration of the OH groups [p.p.m.),
and Φ_0, Φ are the values read from the graphical representation of the absorption band at $\lambda = 2.73 - 2.85\ \mu\text{m}$ (see Figure 120).

Table 29

Composition [atom. %]	Range of spectral transmission [μm]
$Si_{25}As_{25}Te_{50}$	2—9
$Ge_{10}As_{20}Te_{70}$	2—20
$Si_{15}Ge_{10}As_{25}Te_{50}$	2—12.5
$Ge_{30}P_{10}S_{60}$	2—8
$Ge_{40}S_{60}$	0.9—12
$Ge_{28}Sb_{12}Se_{60}$	1—15
$Ge_{33}As_{12}Se_{55}$	1—16
$As_{50}S_{20}Se_{30}$	1—13
$As_{50}S_{20}Se_{20}Te_{10}$	1—13
$As_{35}S_{10}Se_{35}Te_{20}$	1—12
$As_{38.7}Se_{61.3}$	1—15
As_8Se_{92}	1—19
$As_{40}S_{60}$	1—11

In contrast to other types of glasses, the chalcogenide glasses are transparent in the infra-red region of the spectrum. Depending upon the composition of the glass, the region of transmission ranges between the wavelengths of 0.9 and 20 μm. Table 29 lists (Hilton [14]) various types of composition of chalcogenide glasses and the corresponding ranges of spectral transmission of infra-red radiation.

162

Figure 121 presents curves of spectral transmission in chalcogenide and other types of glasses.

The shape of the spectral transmission curve (absorption) is influenced by the presence of oxides or OH groups. Selecting an appropriate production technology is therefore of utmost importance. The chalcogenide glasses have recently received particular attention from a number of researchers and technologists. For more detailed information and data on this subject the reader is therefore referred to references [15, 16, and 17].

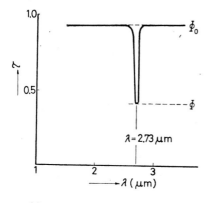

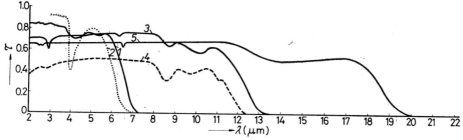

Fig. 120 Schematic diagram of the Φ_0 and Φ values used to calculate the concentration of the OH groups in quartz glass.

Fig. 121 Spectral transmission in glasses: *1* — PbO—GeO$_2$ glass; *2* — CaO—Al$_2$O$_3$—SiO$_2$ glass; *3* — As$_2$S$_3$ glass; *4* — As$_{2.75}$Te$_{0.25}$ glass; *5* — As$_2$Se$_3$.

4.3.4 Dependence of absorption upon temperature

Elevated temperatures are accompanied by variations in the curves of spectral transmission due to the loosening of the structure: micro-heterogeneous scattering centres may also be established and the valency equilibrium of the colouring ions shifted. Figure 122 presents curves of spectral absorption of infra-red radiation in window glass at various temperatures (after Grove [18]). This spectral region is most significant as far as the evaluation of the propagation of thermal radiation through glass is concerned.

If the glass contains colouring ions, e.g., transition elements, the electron transitions ($d-d$ transitions) will be weak at normal room temperature. In the near

163

infra-red region temperature exerts influence upon the vibrations of the complexes (according to [19]), so that, for example, the Fe^{2+} ion manifests pronounced absorption in the near infra-red region of the spectrum. The ligand field is so strong that the Fe^{2+} ion and the ligands will form high-spin complexes, in which transitions then take place. This leads to an excited state in which the equilibrium

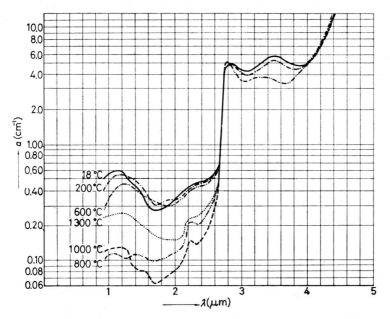

Fig. 122 Spectral absorption of window glass in the infra-red region of the spectrum related to temperature.

inter-nuclear spacings are larger than those in the ground state; and this fact then obviously gives rise to asymmetric vibrations active in the near infra-red region of the spectrum. This interaction absorbs the energy emitted, which results in a fall-off in the transmission of infra-red radiation by the glass. Figure 123 illustrates the absorption curves for the SVAR welder's glass containing a large share of iron monoxide and iron trioxide.

4.3.5 Dependence of absorption upon the physical state of the glass

Glasses undergo structural changes in the transformation range. These changes are reversible and they depend upon the temperature and the length of the period of the thermal treatment. The changes of the physical properties can also be accompanied by variations in the spectral characteristics of coloured glasses.

164

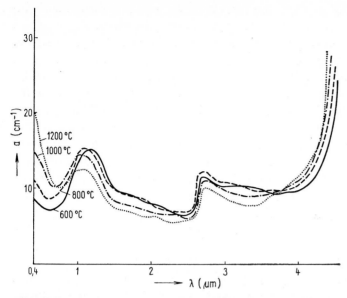

Fig. 123 Spectral absorption of the Svar welder's glass in the infra-red region of the spectrum related to temperature.

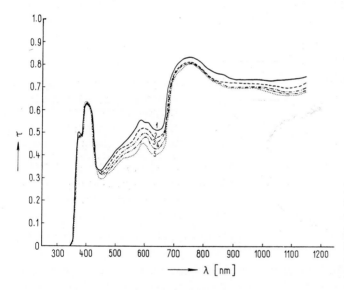

Fig. 124 Spectral characteristics of the Rosal sun glass thermally untreated (*1*), and thermally treated at 490 °C for 10 min (*2*), for 30 min (*3*), for 60 min (*4*), and for 300 min (*5*).

For instance, nickel forms two different coordinations, [4] and [6], in glass (according to Weyl and Thömen [20]). The coordination [6] nickel will colour the glass yellow, the absorption band being placed at $\lambda = 440$ nm. The coordination [4] nickel, however, will stain the glass purple, and the absorption band will be in the visible region of the spectrum.

Both these colour centres are always present in the glass, but their ratio is influenced, among other factors, by the temperature and the length of the period of heat treatment within the range of the temperature of the transformation interval. When the temperature rises, coordination [4] will predominate, and an abrupt quench can preserve this state. These problems have been discussed by Fanderlik [21], who studied the phenomena of transformation. In Figure 124 we present the results of the studies of the changes in the spectral characteristics of Rosal sun glass (which contains, among other colourants, nickel) heated up to temperatures within the range of the transformation temperature (after Fanderlik [22]).

With prolonged thermal treatment, the absorption band will gradually deepen, particularly at the $\lambda = 640$ nm wavelength. In conformity with the theory, the purple tint will grow more intense, corresponding to the prevailing [4] coordination of nickel.

4.3.6   Methods of measurement and apparatus

Spectral transmittance, τ_λ, or absorptance, α_λ, are measured by *spectrophotometers* from Hilger Uvispek, Optica NIDR, Beckman DK-2A, Perkin-Elmer 330 and 350, Cary 14, 14R, 14H and others. Spectrophotometers of various design are used for determining the ratio of two spectral radiation quantities within the range of the ultra-violet, the visible, and the infra-red regions of the spectrum

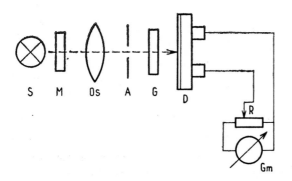

S   M   Os   A   G   D

Fig. 125 Schematic diagram of a single-beam spectrophotometer:
s — light source;
M — monochromator;
Os — optical system;
A — aperture; G — glass specimen; D — selenium photocell; R — resistor;
Gm — galvanometer.

viz. within the optical range. Spectrophotometers with photoelectric indication are most widely used both in research into glass and in the glass industry. For measurements of materials with extremely low optical loss, laser calorimetric techniques have recently been developed [61]. We shall deal with spectrophotometers in more detail, omitting however descriptions of visual and photographic instruments.

A spectrophotometer consists of four principal parts:

(a) a light source,

(b) a monochromator,

(c) an optical system, and

166

(d) a device indicating the intensity of the radiant flux that emerges from the specimen of the glass under examination, comparing it with a standard and recording it automatically.

Two types of spectrophotometers are invariably made use of: the single-beam type and, in the majority of cases, the double-beam type.

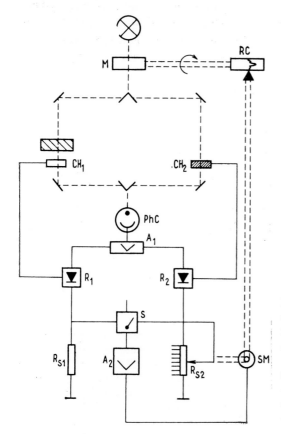

Fig. 126 Schematic diagram of an electrically compensated double-beam recording spectrophotometer: s — light source; M — monochromator; CH_1, CH_2 — choppers; PhC — photoelectric cell; A_1, A_2 — amplifiers; R_1, R_2 — rectifiers; RC — recording; S — switch; SM — servomotor; Rs_1, Rs_2 — resistors.

With the *single-beam type of spectrophotometer* (see diagram in Figure 125) two measurements must always be made for a single wavelength of the luminous flux.

For the first measurement, the specimen of the glass to be measured is not placed in the measuring position and the detector is set for 100 %. To take the second measurement, we insert the glass specimen and take the reading of the percentage intensity transmitted. The intensity of the monochromatic radiant flux is then measured for various wavelengths.

The principle of a *double-beam spectrophotometer* is the equalization of the intensities of two radiant fluxes. As soon as this is achieved, the photoelectric currents of the two detectors become equal, and the galvanometer will not record

167

any deviation. The equalization is attained through compensation. The most widely used is electric and optical compensation. Electric compensation makes use of a resistor or compensation voltage; optical compensation, on the other hand, is achieved with the help of a neutrally grey piece of glass cut in a wedge-like shape and which manifests linear transmittance ranging from zero to 100 %.

In a double-beam spectrophotometer the modulated luminous flux incident upon the detector gives rise to a pulsating photoelectric current, which can be amplified. With the help of servomechanisms governed by potentiometers or the wedge, spectra are automatically recorded as a function of wavelength.

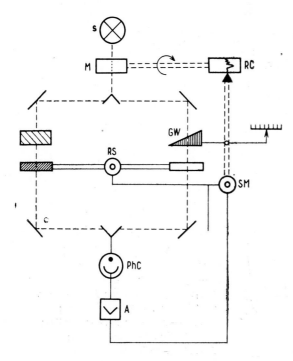

Fig. 127 Schematic diagram of an optically compensated double-beam recording spectrophotometer: s — light source; M — monochromator; RS — rotating sector; GW — wedge; RC — recording; PhC — photoelectric cell; SM — servomotor; A — amplifier.

Figure 126 presents a schematic diagram of a double-beam recording spectrophotometer with a single photoelectric cell and electric compensation. Figure 127 is a diagram of a double-beam spectrophotometer with a photoelectric cell and optical compensation (after [23]).

We shall now deal briefly with the principal parts of a spectrophotometer.

Sources of light

The source of radiant energy must produce continuous radiation of satisfactory intensity and of the required waveband. The most widely used source of ultraviolet radiation is a hydrogen, or deuterium, discharge lamp emitting continuous radiation of about 200—380 nm. For this region of radiation use can also be made of a high-pressure xenon lamp.

For the visible region of the spectrum and the adjoining region of infra-red radiation, use is made of the sources based on the principle of the emission of radiation by a solid body, viz. various types of incandescent lamps with heated coiled tungsten filaments or strips. For the infra-red region of radiation, a heated silit rod placed in a water-cooled envelope is currently used.

The sources of radiation are dealt with in more detail in Section 3.8 and in references [23, 24 and 25].

Monochromators

Monochromators are instruments capable of isolating radiation of a very narrow spectral region. The most currently used are prism monochromators, grating monochromators, and combined monochromators. They operate as optical dispersion instruments and consist of an entrance slit, imaging optical device, exit slit, and a prism or a grating (or a combination of the latter two). A diagram of a prism monochromator is presented in Figure 128, a grating monochromator is sketched schematically in Figure 129.

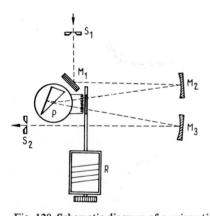

Fig. 128 Schematic diagram of a prismatic monochromator:
S_1, S_2 — slits; M_1, M_2, M_3 — mirrors; P — prism; R — prism rotating device.

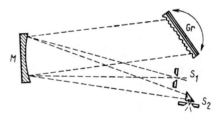

Fig. 129 Schematic diagram of a grating monochromator:
S_1, S_2 — slits, M — mirror; Gr — grating.

The purity of the spectrum depends upon the width of the entrance slit. If the illumination is by means of a line spectrum source and the entrance slit is narrow, the images of the spectrum lines will not overlap. If, however, the spectrum of the source is continuous, the images of the entrance slit pertaining to neighbouring wavelengths will overlap in the spectrum, and the radiation emerging from the exit slit will in part be impure, however narrow the exit slit may be.

Curvature of the spectrum lines is caused by the fact that the entrance slit is not infinitely short: its length is finite. But curving the entrance or exit slits will purify the output spectrum. Monochromatic radition contains about 1 % scattered radiation of adjacent wavelengths. This scatter can be reduced with the help of two dispersing prisms or a prism-and-grating combination.

Dispersion of light by prism

The dispersion of light by a prism is schematically shown in Figure 12. Let us now consider two rays of light the wavelengths of which differ by $\Delta\lambda$. On leaving the prism, these rays will deviate from each other by an angle $\Delta\delta$. If Δn (the change in the refractive index) corresponds to the $\Delta\lambda$ change in the wavelength, then $\dfrac{\Delta\delta}{\Delta\lambda}$ is the angular dispersion of the prism

and

$$\frac{\Delta\delta}{\Delta\lambda} = \frac{2\sin\dfrac{\varphi}{2}}{\sqrt{1 - n^2\sin^2\dfrac{\varphi}{2}}} \cdot \frac{\Delta n}{\Delta\lambda}. \tag{206}$$

Thus, if a prism is used the refracting angle of which is $\varphi = 60°$, then

$$\frac{\Delta\delta}{\Delta\lambda} = \frac{2}{\sqrt{4 - n^2}} \cdot \frac{\Delta n}{\Delta\lambda}, \tag{207}$$

where $\dfrac{\Delta n}{\Delta\lambda}$ is the characteristic dispersion of the prism.

For the region from 0.4 to 2.8 µm, glass prisms are currently used. For the region 0.25 to 3.0 µm, use is made of prisms of pure quartz glass, and for the range of the wavelengths of 2.5 to 16 µm, the most suitable are the NaCl prisms. For the wavebands of 2.0 to 5.5 µm and 15.4 to 25 µm lithium fluoride and potassium bromide prisms, respectively, are also widely used.

Other important aspects of a dispersing prism are: the angular width of the spectrum, the resolving power etc.

For more detailed information the reader is referred to references [24, 25 and 26].

Diffraction of light by grating

A grating is formed by a number of parallel lines, N, of width b, spaced at constant distance a, which is the so-called grating constant. If a beam of parallel light of wavelength λ meets the grating at angle ε (see Figure 130), then, according to Huygens' principle, each point of the ruled line will become the centre of a light disturbance propagated in all directions [25].

At points *I, II, III, IV* the phase of the wave motion is the same for rays *1, 2, 3, 4;* and similarly at points *I', II', III', IV'* for the rays *1', 2', 3', 4'*.

The light propaganting along the path *2B2'* will travel a distance that is longer than the distances of the paths *IAI'* and *IIBII'*. And the light following the paths *3C3'* and *4B4'* will thus cover distances that are longer by sections *IIICIII'* and *IVDIV'*, respectively.

170

Then

$$\overline{IIICIII'} = \overline{2IIBII'},$$
$$\overline{IVDIV'} = \overline{3IIBII'}, \tag{208}$$
$$\overline{IIBII'} = a(\sin \varepsilon + \sin \alpha).$$

Since the light in the *1'*, *2'*, *3'*, ... parallel beam is coherent, the individual rays will interfere with one another. In certain α directions the light will thus be enhanced or attenuated by interference. In this case the most significant will be those α directions in which the reinforcement is maximal. It is therefore essential that the $a(\sin \varepsilon + \sin \alpha)$ path difference should equal an integer multiple of the wavelength, viz.

$$a(\sin \varepsilon + \sin \alpha) = m\lambda, \tag{209}$$

where *m* stands for 1, 2, 3, ... (order of spectrum).

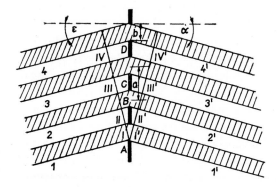

Fig. 130 Diffraction by a grating.

If the angle ε is constant, the angle α will depend upon the wavelength, λ. Thus, if a beam of white light meets the grating, the latter will assume the role of a dispersing system; for the dispersion of this system we can thus write

$$\frac{d\alpha}{d\lambda} = \frac{\Delta\alpha}{\Delta\lambda} = \frac{m}{a \cos \alpha}. \tag{210}$$

For α = 0 the angular dispersion of the grating will be minimal and independent of the wavelength. The angular distance between two spectral lines will thus be equal, if the difference of their wavelengths is the same, regardless of the values of those wavelengths.

The fact that dispersion is independent of the wavelength is the principal characteristic differentiating a grating from a prism.

A grating is further characterized by the angular distance between two adjoining lines, the angular width of the spectrum, the resolving power of the grating, etc.

For more data on this subject the reader is referred to references [23, 24, 25, and 26].

Radiation energy is converted into another kind of energy with the help of photoelectric cells, multipliers, photovoltaic or photoconductive cells (converting radiant energy into electric energy), thermo-electric couples, bolometers (converting radiant energy into thermal and electric energies), and photographic layers (converting radiant energy into photochemical energy) [27].

The detector circuit consists of a detector, an amplifier, and a recorder, which converts the quantity transformed by the detector into a recordable quantity. For accurate measurement, account must be taken of the fact that the detector is noisy; the noise can be suppressed, or separated from the incoming signal, in a number of ways.

The properties of a detector are characterized by the so-called photometric quantity with the help of which we define the flux, the luminance, and the exposure. These properties are also characterized by sensitivity, viz. the outgoing signal relative to the incoming unit signal. Spectral sensitivity is also used for determining the relative distribution of sensitivity with respect to the incident wavelengths; this should ensure that the right range of operation of the detector is selected. Further characteristics are the so-called minimum detectable incoming signal and the superficial resolving capacity.

Photoelectric cells and multipliers are based on the principle of the external photoelectric effect. The vacuum photoelectric cells operate on the principle of the electrons released from the photocathode being accelerated towards the anode. In gas phototubes, on the other hand, the electrons impinging upon the molecules of gas ionize these molecules, so that the number of the electrons incident upon the anode will increase.

In multipliers the electrons released from the cathode impinge upon the first amplifying electrode, where secondary emission will multiply their number. They then meet the second, third, and further amplifying electrodes, their numbers increasing, so that the anode receives electrons whose number is higher by several orders than the number of electrons leaving the cathode. Only in the vacuum photoelectric cells and in multipliers is the photoelectric current proportional to radiant flux, but not in gas phototubes to any major extent.

The principle on which the *photoelectric cells* operate is the photovoltaic effect. The most widely used is the selenium cell, which is basically a plate coated with a layer, viz. a photovoltaic layer, of semiconductive selenium. Applied upon this layer is a thin coat of light-transmitting metal. If the outer circuit of the cell is shorted, the system is traversed by a photoelectric current, which is dependent upon both the illumination and the area of the effective surface of the cell. The higher the resistance of the cell, the more will the linear dependence of the current upon illumination be affected. But with these cells fatigue and delayed action are pretty common features.

The *photoconductive cells* operate on the principle of the internal photoelectric effect. Use is mostly made of thallium monosulphide, lead telluride, and lead sulphide. Germanium detectors with the mixtures of gallium, copper, silver, etc. have been introduced only recently.

The *thermoelectric couples* and *thermopiles* are based upon the exploitation of the thermoelectric effects. They are made up of two different metals, which produce electric current of maximum possible voltage if heated at the junction. Thermopiles consist of several thermoelectric cells coupled in series; thermoelectric voltage is proportional to thermal difference between the so-called hot junction and the cold ends of the wires. Thermopiles manifest considerable delay of action the measure of which is a time constant.

The operation of a bolometer is based on the dependence of the resistance of the Si, Ni, Bi metal conductors or CuO, MnO, CoO semiconductors upon temperature. They manifest a much shorter time constant than the thermoelectric couples.

Measurement of spectral transmittance at temperatures above the softening point

Spectral transmittance of glasses at elevated temperatures can be measured in various ways, the most important being the method of measurement making use of a *platinum cell with sapphire windows*, or a *platinum crucible* and the reflection of a beam of light from the polished bottom of the crucible.

The former is usually applied in the measurement of the spectral transmittance of glasses at elevated temperatures in the infra-red region of the spectrum, which is of utmost importance for the radiant transfer of heat. The values recorded can be employed in calculations of radiant heat conductivity. The measurement makes use of a special platinum cell with sapphire windows placed in an electric oven. The single-beam infra-red spectrophotometer used [28] has an extended optical path and the selective modulation at 400 Hz makes it possible to neutralize the radiation of the oven. It is essential that the effect of the thermal expansion of the platinum cell upon the thickness of the layer of molten glass and the reflection of the traversing radiation should be neutralized. In Figure 123 we present the results for the Svar welder's glass obtained using this method.

The method utilizing reflection of light from the polished bottom of a platinum crucible is also widely used. The beam of light emerging from the source is modulated by a chopper and focused by a suitable optical system onto the bottom of the platinum crucible at a low angle of incidence. The reflected beam then traverses the molten glass layer and is optically concentrated on the entrance slit of the monochromator. Emerging from the monochromator, the beam meets the detector. This method is applied in the measurement to be made in either the visible or the near infra-red region of the spectrum up to approximately 2.5 μm.

Measurement of spectral transmittance of glasses
at liquid nitrogen temperature

The measurement is carried out with conventional spectrophotometers. A cryostat must, however, be applied. This is inserted, together with the specimen of the glass to be measured, into the spectrophotometer. The required temperature (−195 °C) is then reached in the cryostat with the help of liquid nitrogen (see Figure 131).

The specimen is placed in a tank with tiny windows of optical quartz glass. In order that the windows remain clear, the tank is evacuated. In the course of the measurement it is, however, necessary to keep adding further quantities of liquid

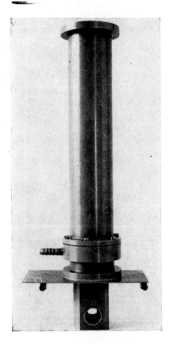

Fig. 131 Cryostat.

nitrogen, because liquid nitrogen evaporates rapidly at room temperatures. The method is particularly interesting owing to the fact that in some cases the reversible thermal effects will get "frozen", so that measurements will be unaffected by any ambient temperatures.

Measurement of spectral transmittance of powdered
glasses in the infra-red region of the spectrum

The spectral transmittance of glasses, with the exception of the chalcogenide glasses, is limited to approximately the 5 μm wavelength. The study of the structure of glasses, however, also requires measurement in the infra-red region. The specimens of the glass to be measured are prepared in such a way that the glass is

174

reduced to fine-grained powder capable of passing through a screen of 1460 meshes per one square centimeter. The powder is then mixed with potassium bromide and compacted into tablets. The concentration of the glass to be measured is approximately $0.1-1.0\%$ of the potassium bromide content. The tablets are then evacuated (1 999 Pa) by pressing, the pressure required equalling approximately 735 MPa. For materials exhibiting a high refractive index the usual mixtures are thallium monobromide (for the transmission of radiation of the range of $1-25$ μm), caesium bromide, caesium iodide, etc. This method is described in detail in reference [29] and is much more convenient than the method in which specimens are prepared by ultra-microtoming, which very often poses great problems.

Measurement of spectral transmittance of glasses by the method
of the absorption of reflected radiation

In certain cases of the measurement of the spectral transmittance of glasses a most advantageous method seems to be the one making use of the application of a thin glass layer onto a metal plate [24]. One part (Φ') of the incident radiant flux, Φ_0, is reflected by the surface of the glass layer, the other part, Φ, will pass through the layer and will be reflected by the surface of the metal plate, traverse the glass layer for the second time, and leave the system. But part of the radiant flux is returned to the surface of the metal plate by the air/glass interface, is reflected again and will emerge from the system in the form of radiant flux Φ''. We thus have the case of multiple reflection at the interfaces of the system. The intensities of the individual radiant flows can be expressed as

$$\Phi' = \varrho\Phi_0,$$
$$\Phi = (1 - \varrho)^2 . \Phi_0 . 10^{-alc},$$
$$\Phi'' = (1 - \varrho)^3 . \Phi_0 . 10^{-2alc}, \tag{211}$$

where ϱ is the reflectance,
$\quad a \quad$ the linear absorption coefficient,
$\quad l \quad$ the thickness of the glass layer,
and $\quad c \quad$ the concentration of the absorption centres.

Owing to the phase shift the reflected radiant flux is amplified, or attenuated, at various wavenumbers. Disregarding Φ'' and the higher radiant flows, we can write for the total intensity

$$\Phi_e = \Phi + \Phi' + \Phi'' + 2(\sqrt{\Phi\Phi'} - \sqrt{\Phi\Phi''}) \cos \vartheta - 2\sqrt{\Phi'\Phi''} \cos 2\vartheta \tag{212}$$

where

$$\vartheta = \frac{4\pi l(n^2 - \sin^2\Theta)^{1/2}}{\lambda}, \tag{213}$$

where n is the index of refraction,
and $\quad \Theta \quad$ the angle of incidence.

Interferential absorption bands can thus be observed in the reflection spectra. Of course, the intensity of the radiation is always higher with reflection methods than with methods based on transmission.

Measurement of colour

The colour quality of glasses, particularly of coloured glasses, is measured by colorimeters. For instance, the J-40 colorimeter [30] has four filters measuring the magnitudes of three tristimulus values of the light emerging from the specimen measured. The illumination is by CIE standard achromatic light C of equivalent colour temperature 6 500 K (normal magnesium oxide). Filters X_1, X_2, Y and Z are used and the chromaticity coordinates are calculated from the values recorded, viz.

$$x = \frac{X}{X + Y + Z}; \quad y = \frac{Y}{X + Y + Z}; \quad (214)$$

$$z = \frac{Z}{X + Y + Z}; \quad X = X_1 + X_2.$$

The Y value also gives the total transmission of the light. With perfectly prepared specimens the accurancy of measurement equals ± 0.001 for chromaticity coordinates x, y, z.

4.3.7 Calculations

The metod calculating internal transmittance, τ_i, from the spectral transmission factor, τ, has been discussed in Section 4.1.9. Apart from the losses due to reflection, the values of the spectral transmission also depend on the thickness of the glass. It is therefore common that the recorded values are given for a thickness that has, for some reason, been properly adjusted to suit the measurement.

The thickness of the glass exerts considerable influence on the spectral characteristics, particularly of coloured glasses. We give below an example of the calculation of internal transmittance, τ_i, of an l_2 thick specimen using the values recorded for a specimen the thickness of which is l_1.

Example 1

Let us convert the values of the transmission factor recorded for a 2 mm thick Greenal glass specimen into those corresponding to a 4 mm thickness ($l_2 = 4$ mm), the refractive index of the glass being $n_d = 1.518$.

The values of the spectral transmission factor recorded and expressed as τ. 100 [%] for a 2-mm-thick glass ($l_1 = 2$ mm) are quoted in the second column of Table 30. In the third column are values corrected for losses due to reflection, i.e. values of the so-called internal transmittance τ_i. 100 [%] for the $l_1 = 2$ mm thickness. This procedure is, of course, identical with the one suggested in Table 12.

Table 30

λ [nm]	$\tau \cdot 100$ [%]	$\tau_t \cdot 100$ [%]	D_{i1}	D_{i2}	$\tau_t \cdot 100$ [%]
340	12.5	13.6	0.866	1.732	1.8
360	55.0	59.8	0.223	0.446	36.0
380	56.5	61.5	0.211	0.422	38.0
400	69.5	75.6	1.121	0.242	57.0
420	48.0	52.2	0.284	0.584	27.0
440	38.0	41.9	0.378	0.756	17,5
460	40.5	44.1	0.356	0.712	19.5
480	46.0	50.1	0.300	0.600	25.0
500	54.0	58.8	0.231	0.462	35,0
520	60.0	65.3	0.185	0.370	42.5
540	61.0	66.4	0.177	0.354	44.0
560	60.5	65.8	0.182	0.364	43.5
580	59.5	64.7	0.180	0.366	42.5
600	56.0	60.9	0.215	0.430	37.0
620	49.0	53.3	0.273	0.546	28.5
640	47.0	51.1	0.292	0.584	26.0
660	48.5	52.8	0.277	0.554	28.0
680	54.5	59.3	0.227	0.454	35.5
700	55.0	59.8	0.223	0.443	36.0

The next step is to calculate the internal transmission density, D_{i1}, for the thickness l_1 (column four) and, using Lambert's equation, then obtain the value D_{i2}. The latter is quoted in the fifth column of Table 30.

$$\frac{D_{i1}}{D_{i2}} = \frac{l_1}{l_2}. \tag{215}$$

D_{i2} may also be derived from the equation

$$D_{i2} = \log \frac{\Phi_{tr}}{\Phi}, \tag{216}$$

or may be read from Table 31. The results are given in the sixth column of Table 30.

Example 2

For certain cases of the calculation of spectral transmittance of optical glasses a method has been suggested for calculating the absorption factor for the 400 – 750 nm range of the visible region of the spectrum from measured values of the spectral transmission factor.

Let us calculate the values of the absorption factor, α_λ, of a 10-mm-thick glass from the values of the spectral transmission factor recorded for an 83 mm thickness.

Table 31

$\tau \cdot 100\ [\%]$	D_i	$\tau \cdot 100\ [\%]$	D_i	$\tau \cdot 100\ [\%]$	D_i
0.5	2.301	24.5	0.611	48.5	0.314
1.0	2.000	25.0	0.602	49.0	0.310
1.5	1.824	25.5	0.593	49.5	0.305
2.0	1.699	26.0	0.585	50.0	0.301
2.5	1.602	26.5	0.577	50.5	0.297
3.0	1.523	27.0	0.569	51.0	0.292
3.5	1.456	27.5	0.561	51.5	0.288
4.0	1.398	28.0	0.553	52.0	0,284
4.5	1.347	28.5	0.545	52.5	0.280
5.0	1.301	29.0	0.538	53.0	0.276
5.5	1.260	29.5	0.530	53.5	0.272
6.0	1.222	30.0	0.523	54.0	0.268
6.5	1.187	30.5	0.516	54.5	0.264
7.0	1.155	31.0	0.509	55.0	0.260
7.5	1.125	31.5	0.502	55.5	0.256
8.0	1.097	32.0	0.495	56.0	0.252
8.5	1.071	32.5	0.488	56.5	0.248
9.0	1.046	33.0	0.481	57.0	0.244
9.5	1.022	33.5	0.475	57.5	0.240
10.0	1.000	34.0	0.468	58.0	0.237
10.5	0.979	34.5	0.462	58.5	0.233
11.0	0.959	35.0	0.456	59.0	0.229
11.5	0.939	35.5	0.450	59.5	0.225
12.0	0.921	36.0	0.444	60.0	0.222
12.5	0.903	36.5	0.438	60.5	0.218
13.0	0.886	37.0	0.432	61.0	0.215
13.5	0.870	37.5	0.426	61.5	0.211
14.0	0.854	38.0	0.421	62.0	0.208
14.5	0.839	38.5	0.414	62.5	0.204
15.0	0.824	39.0	0.409	63.0	0.201
15.5	0.809	39.5	0.403	63.5	0.197
16.0	0.796	40.0	0.398	64.0	0.194
16.5	0.782	40.5	0.392	64.5	0.190
17.0	0.770	41.0	0.387	65.0	0.187
17.5	0.757	41.5	0.382	65.5	0.184
18.0	0.745	42.0	0.377	66.0	0.180
18.5	0.733	42.5	0.372	66.5	0.177
19.0	0.721	43.0	0.366	67.0	0.174
19.5	0.710	43.5	0.361	67.5	0.171
20.0	0.699	44.0	0.356	68.0	0.167
20.5	0.688	44.5	0.352	68.5	0.164
21.0	0.678	45.0	0.347	69.0	0.161
21.5	0.667	45.5	0.342	69.5	0.158
22.0	0.657	46.0	0.337	70.0	0.155

Table 31 (cont.)

$\tau \cdot 100$ [%]	D_i	$\tau \cdot 100$ [%]	D_i	$\tau \cdot 100$ [%]	D_i
22.5	0.648	46.5	0.332	70.5	0.152
23.0	0.638	47.0	0.328	71.0	0.149
23.5	0.629	47.5	0.323	71.5	0.146
24.0	0.620	48.0	0.319	72.0	0.143
72.5	0.140	82.0	0.086	91.5	0.039
73.0	0.137	82.5	0.084	92.0	0.036
73.5	0.134	83.0	0.081	92.5	0.034
74.0	0.131	83.5	0.078	93.0	0.032
74.5	0.128	84.0	0.077	93.5	0.029
75.0	0.125	84.5	0.073	94.0	0.027
75.5	0.122	85.0	0.071	94.5	0.025
76.0	0.119	85.5	0.068	95.0	0.022
76.5	0.116	86.0	0.066	95.5	0.020
77.0	0.113	86.5	0.063	96.0	0.018
77.5	0.111	87.0	0.061	96.5	0.016
78.0	0.108	87.5	0.058	97.0	0.013
78.5	0.105	88.0	0.056	97.5	0.011
79.0	0.102	88.5	0.053	98.0	0.009
79.5	0.100	89.0	0.051	98.5	0.007
80.0	0.097	89.5	0.048	99.0	0.004
80.5	0.094	90.0	0.046	99.5	0.002
81.0	0.092	90.5	0.043	100.0	0.000
81.5	0.089	91.0	0.041		

In the second column of Table 32 we quote the $\tau \cdot 100$ [%] values recorded. These values must first be multiplied by the factors of column three of Table 32. The results are presented in column four.

The spectral transmission factor, $\tau_i \cdot 100$ [%], free from losses due to reflection, is calculated for the 83-mm-thick glass for the quoted range of the wavelengths and for reflection factor $R = 0.875$.

$$\frac{7\,080\,699.0}{0.875 \cdot 10\,000\,000} = 80.92 \quad [\%]. \tag{217}$$

The values of the spectral transmission factor of the 83-mm-thick glass are then converted into those corresponding to the 10 mm thickness, viz.

$$\frac{D_{i2}}{D_{i1}} = \frac{l_2}{l_1} = \frac{D_{i2}}{0.0919} = \frac{10}{83}, \tag{218}$$

where the values of D_{i1} and D_{i2} can be calculated or read from Table 31. Thus, if the calculated value D_{i2} equals 0.01108, then $\tau_{i2} \cdot 100 = 97.5\%$. If $\alpha_{i2} + \tau_{i2} = 1$, then $\alpha_{i2} = 0.025$.

Table 32

λ [nm]	τ . 100 [%]	Factor [f]	τ . 100 . f
400	37.5	1	37.5
410	37.5	2	75.0
420	44.0	8	352.0
430	48.5	27	1 309.5
440	49.0	61	2 989.0
450	52.5	117	6 142.5
460	57.0	210	11 970.0
470	59.0	362	21 358.0
480	61.5	622	38 253.0
490	65.0	1 039	67 535.0
500	67.5	1 792	120 960.0
510	68.0	3 080	209 440.0
520	68.5	4 771	326 813.5
530	70.0	6 322	442 540.0
540	71.0	7 600	539 600.0
550	71.5	8 568	612 612.0
560	71.5	9 222	659 373.0
570	70.5	9 457	666 718.5
580	70.0	9 228	645 960.0
590	71.5	8 540	610 610.0
600	70.0	7 547	528 290.0
610	71.5	6 356	454 454.0
620	73.5	5 071	372 718.5
630	73.5	3 704	272 244.0
640	74.0	2 562	189 588.0
650	74.0	1 637	121 138.0
660	74.5	972	72 414.0
670	75.0	530	39 750.0
680	76.0	292	22 192.0
690	76.5	146	11 169.0
700	77.5	75	5 812.5
750	79.5	79	6 280.5

In a number of cases it may be advantageous to convert the values for one thickness into those for another with the help of Vaughan's calculator [32]. The calculation will be accurate provided the value of the refractive index of the glass ranges between about 1.50 and 1.52.

Another possibility is to use the nomograph described in [1].

Example 3

The *chromaticity coordinates* can be very accurately calculated from the curve of spectral transmission. The method is discussed in reference [49]. Use is made of the relationships

$$X = \int E\overline{X}\,d\lambda; \qquad Y = \int E\overline{Y}\,d\lambda; \tag{219}$$
$$Z = \int E\overline{Z}\,d\lambda,$$

where E is the radiation energy in the $d\lambda$ waveband,

$\overline{X}$, $\overline{Y}$, $\overline{Z}$ three wavelength functions defining the qualities of the standard observer, adopted by CIE, given in Table 33.

and X, Y, Z the values for the equienergy source.

Table 33

λ [nm]	$\overline{X}$	$\overline{Y}$	$\overline{Z}$	λ [nm]	$\overline{X}$	$\overline{Y}$	$\overline{Z}$
380	0.0014	0.0000	0.0055	580	0.9163	0.8700	0.0017
390	0.0042	0.0001	0.0201	590	1.0263	0.7570	0.0011
400	0.0143	0.0004	0.0679	600	1.0622	0.6310	0.0008
410	0.0435	0.0012	0.2074	610	1.0026	0.5030	0.0003
420	0.1344	0.0040	0.6476	620	0.8544	0.3810	0.0002
430	0.2839	0.0116	1.3856	630	0.6424	0.2650	0.0000
440	0.3483	0.0230	1.7471	640	0.4479	0.1750	0.0000
450	0.3362	0.0380	1.7721	650	0.2835	0.1070	0.0000
460	0.2908	0.0600	1.6692	660	0.1649	0.0610	0.0000
470	0.1954	0.0910	1.2876	670	0.0874	0.0320	0.0000
480	0.0956	0.1390	0.8130	680	0.0468	0.0170	0.0000
490	0.0320	0.2080	0.4652	690	0.0227	0.0082	0.0000
500	0.0049	0.3230	0.2720	700	0.0114	0.0041	0.0000
510	0.0093	0.5030	0.1586	710	0.0058	0.0021	0.0000
520	0.0633	0.7100	0.0782	720	0.0029	0.0010	0.0000
530	0.1655	0.8620	0.0422	730	0.0014	0.0005	0.0000
540	0.2904	0.9540	0.0203	740	0.0007	0.0003	0.0000
550	0.4334	0.9950	0.0087	750	0.0003	0.0001	0.0000
560	0.5945	0.9950	0.0039	760	0.0002	0.0001	0.0000
570	0.7621	0.9520	0.0021	770	0.0001	0.0000	0.0000

The Y value is simultaneously a function of the relative luminous efficiency, so that it enables direct calculation of total transmittance. For instance, for the signal glasses, where the curve of spectral transmission is relatively simple, the integration can be substituted by a summation at 10 nm intervals, so that for each coordinate we obtain the X, Y, Z values by simply multiplying the transmission factor of the measured glass by the spectral distribution factor for the given equivalent colour temperature of the light source, and by finally adding up all the products. From these values we calculate the chromaticity coordinates, viz.

$$x = \frac{X}{X + Y + Z}; \qquad y = \frac{Y}{X + Y + Z};$$

$$z = \frac{Z}{X + Y + Z}. \tag{220}$$

Let us now calculate the chromaticity coordinates of a very dark black glass for a source of the equivalent colour temperature 1900 K. For the result see Table 34. First, we calculte the X, Y, Z values from the relationships

$$X = \sum_{380}^{780} \tau_\lambda(\mathbf{k}E'_{\lambda, T}\bar{X}\,\Delta\lambda) = \sum_{380}^{780} \tau_\lambda f(x, T),$$

$$Y = \sum_{380}^{780} \tau_\lambda(\mathbf{k}E'_{\lambda, T}\bar{Y}\,\Delta\lambda) = \sum_{380}^{780} \tau_\lambda f(y, T), \tag{221}$$

$$Z = \sum_{380}^{780} \tau_\lambda(\mathbf{k}E'_{\lambda, T}\bar{Z}\,\Delta\lambda) = \sum_{380}^{780} \tau_\lambda f(z, T),$$

where τ_λ is the recorded value of the transmittance factor for λ wavelengths.

Table 34

λ [nm]	τ_λ	$f(x, 1900)$	$\tau_\lambda \cdot f(x, 1900)$ $x = 24\,666.75$	$f(y, 1900)$	$\tau_\lambda \cdot f(y, 1900)$ $y = 9623.59$	$f(z, 1900)$	$\tau_\lambda \cdot f(z, 1900)$ $z = 0.00$
580	0.005	10 040	50.20	9533	47.66	19	0.09
590	0.010	12 874	128.74	9496	94.96	14	0.14
600	0.010	15 168	151.68	9011	90.11	11	0.11
610	0.020	16 192	323.84	8123	162.46	5	0.10
620	0,075	15 537	1165.27	6928	519.67	3	0.15
630	0.250	13 082	3270.50	5396	1349.00	—	
640	0.530	10 168	5389.04	3973	2105.69		
650	0.745	7 143	5321.53	2696	2005.82		
660	0.820	4 591	3764.62	1698	1392.36		
670	0.840	2 677	2248.68	980	823.20		
680	0.845	1 571	1327.49	571	482.49		
690	0.850	832	707.20	301	255.85		
700	0.840	455	382.20	164	137.76		
710	0.830	251	208.33	91	75.53		
720	0.820	136	111.52	47	38.54		
730	0.810	71	57.51	25	20.25		
740	0.800	38	30.40	16	12.80		
750	0.795	17	13.51	8	6.36		
760	0.770	12	9.24	4	3.08		
770	0.750	7	5.25	—	—		

The values of that factor are given in Table 34, second column. Multiplying by factors f_x, f_y, f_z given in the third, fifth, and seventh column of Table 34, we get the values $\tau_\lambda . f_x . 1900$ K, $\tau_\lambda . f_y . 1900$ K, and $\tau_\lambda . f_z . 1900$ K. A simple addition will give us the X, Y, Z values. The factors quoted in Table 34 are based on the factors of Table 33, multiplied, of course, by a constant, which provides us with the possibility of calculating total transmission.

The chromaticity coordinates thus are

$$x = \frac{X}{X + Y + Z} = \frac{24\,666.75}{34\,290.93} = 0.7193,$$

$$y = \frac{Y}{X + Y + Z} = \frac{9623.59}{34\,290.93} = 0.2806, \qquad (222)$$

$$z = \frac{Z}{X + Y + Z} = \frac{0.59}{34\,290.93} = 0.0001.$$

Total transmittance equals

$$\tau = \frac{Y}{100\,000} = \frac{9623.59}{100\,000} = 0.0962.$$

References

[1] J, Kocík, J. Nebřenský and I. Fanderlik, Barvení skla (Colour Generation in Glass). SNTL Prague 1978, 2nd edn.

[2] T. Bates, Ligand Field Theory and Absorption spectra of Transition Metal Ions in Glasses. In: Modern Aspects of the Vitreous State, Vol. 2. Butterworths, London 1963.

[3] C. R. Bamford, The application of ligand field theory to coloured glasses. *Phys. Chem. Glasses*, *3* (1962) pp. 189—202.

[4] W. D. Johnston, Oxidation-reduction equilibria in iron-containing glass. *J. Am. Ceram. Soc.*, *47* (1964) pp. 198—201.

[5] M. V. Garino-Canina, Contribution a la connaissance de la silice vitreuse. Thesis, Univ. de Paris, 1956.

[6] W. A. Weyl, Coloured Glasses. The Society of Glass Technology, Sheffield 1951.

[7] W. Nowotny, Szkla barwne (Coloured Glasses). Arkady, Warsaw 1958.

[8] W. Weyl and E. CH. Marboe, The Constitution of Glasses. A Dynamic Interpretation. Interscience, New York 1962; The Constitution of Glasses, Vol. 2. Interscience, New York 1964.

[9] H. Scholze, Der Einbau Wassers in Gläsern. *Glastechn. Ber.*, *32* (1959) pp. 81—88, 142—152, 278—281, 314—320, 381—386.

[10] H. Scholze, Zur Frage der Unterscheidung zwischen H_2O Molekeln und OH-Gruppen in Gläsern und Mineralen. *Naturwiss.*, *47* (1960) pp. 226—227.

[11] J. Yovanovitch, Quelques resultets concernant l'influence sur la viscosité de la silice vitreuse, de la quantité d'eau quelle contient. *Bul. Soc. Fr. Céram.*, *66* (1962) pp. 25—30.

[12] H. Scholze and H. Franz, Zur Frage der IR-Bande bei 4.25 μm in Gläsern. *Glastechn. Ber.*, *35* (1962) pp. 278—281.

[13] I. Simon, Infra-red studies of glass. In: Modern Aspects of the Vitreous State, Vol. 1. Butterworths, London 1960.

[14] A. R. Hilton, Optical properties of chalcogenide glasses. *J. Non Cryst. Solids*, *2* (1970) pp. 28—39.

[15] N. F. Mott and E. A. Davis, Elektronové procesy v nekrystalických látkách (Electronic Processes in Non-Crystalline Substances). Jednota čs. matematiků a fyziků, Prague 1974.

[16] Amorphous and Liquid Semi-Conductors. International Conference on Amorphous and Liquid Semi-Conductors. Cavendish Laboratory, Cambridge, 1969.

[17] H. Rawson, Inorganic Glass-Formation Systems. Academic Press, London, New York 1967.

[18] F. J. Grove, Spectral transmission of glass at high temperatures and its application to the heat-transfer problems. J. Am. Cer. Soc., 44 (1961) pp. 317—320.

[19] M. Coenen, Durchstrahlung des Glasbades bei Farbgläsern. Glastechn. Ber., 41 (1968) pp. 1—10.

[20] W. Weyl and E. Thömen, Über die Konstitution des Glases. II. Theoretisch Grundlagen der Glasfarbungen. Glastechn. Ber., 11 (1933) pp. 113—120.

[21] M. Fanderlik, Chlazení skla — pochod chemicko asociační (Cooling of glass. A chemical associative process). Sklářské rozhledy, 22 (1946) pp. 14—19.

[22] I. Fanderlik, Studie změn spektrálních charakteristik protislunečních skel typu Rosal v transformačním intervalu (A study of the variations of spectral characteristics of Rosal sun glass in the transformation Range). Sklář a keramik, 12 (1964) pp. 335—339.

[23] M. Malát, Absorpční anorganická fotometrie (Absorption Inorganic Photometry). Academia, Prague 1973.

[24] A. Vaško, Infračervené záření a jeho použití (Infrared Radiation and its Application). SNTL, Prague 1963.

[25] B. Havelka, E. Keprt, and M. Hansa, Spektrální analýza (Spectral Analysis). NČSAV, Prague 1957.

[26] J. Brož et al., Základy fyzikálních měření — II B (Essentials of Physical Measurements). SPN, Prague 1974.

[27] Coll. of authors, Analytická příručka (An Analytic Manual). SNTL, Prague 1972.

[28] A. Blažek, J. Endrýs, V. Tydlitát, J. Kada, and J. Staněk, Messung der Strahlungsfähigkeit von Glas bei Temperaturen bis 1 400 °C. Congrés International du Verre, Versailles 1971, Sect. Al-7, pp. 735—755.

[29] I. Kössler, Kvantitativní infračervená spektrometrická analysa (Quantitative Infra-Red Spectrometric Analysis). SNTL, Prague 1970.

[30] ČSN 70 0304 — Měření barevnosti transparentních skel (Czechoslovak Standard No. 70 0304 — Measurement of Chromaticity of Transparent Glasses). 1968.

[31] I. Fanderlik, Changes in spectral characteristic of Czechoslovak sun glasses of the Greenal type caused by furnace atmosphere. Symposium on Coloured Glass, Jablonec 1965.

[32] T. C. Vaughan, A calculator for computing the spectral transmission of glasses at different thicknesses. Glass Ind., 25 (1944) pp. 259, 278—279.

[33] M. Kříž, Ekvivalentní koncentrace barviv v sodnodraselném skle (Equivalent concentration of colourants in soda potash glass). Symposium on Coloured Glass, Prague 1967, pp. 267—269.

[34] J. Wong and C. A. Angell, Glass Structure by Spectroscopy. Dekker, New York, Basel 1976.

[35] E. N. Condon and G. H. Shortley, Theory of Atomic Spectra. Cambridge University Press 1953.

[36] G. Racah, Calculations on classical field theory II. Phys. Rev., 62 (1942) p. 438. Theory of complex spectra III. 63 (1943) p. 367.

[37] H. Bethe, Termausfspaltung in Kristallen. Ann. Phys., 3 (1929) p. 133.

[38] Y. Tanabe and S. Sugano, On the absorption spectra of complex ions I and II. J. Phys. Soc. Japan, 9 (1954) pp. 753 — 766, 766 — 779.

[39] L. E. Orgel, Introduction to Transition-Metal Chemistry: Ligand Field Theory. Methuen, London 1960.

[40] R. W. Douglas, Absorption in the visible range of wavelengths due to transition metal ions. Symposium on Coloured Glasses, Prague 1967.

184

[41] W. H. Turner, The electron transfer spectra of silicate glasses. Symposium on Coloured Glasses, Prague 1967.

[42] H. R. Philipp, Optical transitions in crystalline and fused quartz. *Solid State Commun., 4* (1966) pp. 73—75.

[43] G. H. Sigel, Ultraviolet Spectra of silicate glasses: a review of some experimental evidence. *J. Non Cryst. Solids, 13* (1973/1974) pp. 373—398.

[44] A. R. Ruffa, Models for electronic processes in SiO_2. *J. Non-Cryst. Solids, 13* (1973/1974) pp. 37—54.

[45] D. Krause, Aussagen der Absorptionsspektroskopie im sichtbaren und UV zur Struktur von Glas. Schott-Forschungsberichte, Jenaer Glaswerk Schott, Mainz, 1973—1974, pp. 188—218.

[46] V. V. Vargin, Proizvodstvo cvetnovo stekla (Production of Coloured Glass). Gizlegprom, Moscow—Leningrad 1940.

[47] N. Neurot, Aussagen der Spektroskopie im nahen und mittleren Infrarot zur Glasstruktur. Schott-Forschungsberichte, Jenaer Glaswerk Schott, Mainz, 1973—1974, pp. 141—187.

[48] A. Vaško, Infra-red Radiation. Iliffe, London, SNTL, Prague 1963.

[49] J. Kocík and J. Nebřenský, Barevná skla pro signalizaci v železniční dopravě a jejich hodnocení (Coloured signal glasses in rail transport. Their evaluation). *Věda a výzkum v prům. sklářském,* Ser. 1, (1956) pp. 43—85.

[50] C. R. Bamford, Colour Generation and Control in Glass. Elsevier, Amsterdam 1977.

[51] H. Scholze, Le Verre. Institut de Verre, Paris 1969.

[52] P. Nath and R. W. Douglas, Cr^{3+} — Cr^{6+} equilibrium in binary alkali silicate glasses. *Phys. Chem. Glasses, 6* (1965) pp. 197—202.

[53] P. Nath, A. Paul, and R. W. Douglas, Physical and chemical estimation of trivalent and hexavalent chromium in glasses. *Phys. Chem. Glasses, 6* (1965) pp. 203—206

[54] A. Paul and R. W. Douglas, Ferrous-ferric equilibrium in binary alkali silicate glasses. *Phys. Chem. Glasses, 6* (1965) pp. 207—211.

[55] R. W. Douglas, P. Nath, and A. Paul, Oxygen ion activity and its influence on the redox equilibrium in glasses. *Phys. Chem. Glasses, 6* (1965) pp. 216—223.

[56] R. J. Edwards, A. Paul, and R. W. Douglas, Spectroscopy and oxidation-reduction of iron and copper in Na_2O—PbO—SiO_2 glasses. *Phys. Chem. Glasses, 13* (1972) pp. 131—136.

[57] R. J. Edwards, A. Paul, and R. W. Douglas, Spectroscopy and oxidation-reduction of iron in $Mo.P_2O_5$ glasses. *Phys. Chem. Glasses, 13* (1972) pp. 137—143.

[58] A. Paul, Iron-selenium black glass. *Phys. Chem. Glasses, 14* (1973) pp. 96—100.

[59] M. Horák and A. Vítek, Interpretation and Processing of Vibrational Spectra. Elsevier, Amsterdam, Oxford, New York, 1978.

[60] B. D. Mc Swain, N. F. Borrelli, and G. J. Su, The effect of composition and temperature on the ultraviolet absorption of glass. *Phys. Chem. Glasses, 4* (1963) pp. 1—10.

[61] M. Tomozawa, R. M. Doremus, Treatise on materials science and technology, Vol. 12. Glass I.: Interaction with Electromagnetic radiation. Optical Absorption of glasses (G. H. Sigel). Academie press, New York, San Francisco 1977, pp 5—89.

4.4   SCATTERING OF RADIATION BY GLASSES

4.4.1   Introduction

Methods based on the scattering of light have been gaining ground in research into the microheterogeneous nature of glasses in the past few years. From the results of measurements of the intensity of the light scattered by the

185

microheterogeneous scattering centres, we can draw conclusions as to the magnitude and the character of the scattering centres. Various stages of phase separation, nucleation, and crystallization can be followed and colloidal particles of metals, gas bubbles and other heterogeneities examined whose indices of refraction are different from the refractive index of the surrounding glass medium.

The theoretical foundations of the interpretation of the effect of light scatter, which was observed by Tyndall, were laid by Rayleigh. Rayleigh derived an important relationship according to which the intensity of the scattered light is inversely proportional to the 4th power of the wavelength of that light.

Fluctuation theory then supplied an explanation of the scattering of light in liquids. Mie has worked out a theory of the scattering of light by larger scattering centres relating it to the wavelength of the light, and Gans studied the scatter of light in solids.

4.4.2   Theory

The Tyndall effect

Observing light passing through microheterogeneous colloidal systems Tyndall [1] discovered scatter at an angle perpendicular to the incident light, which manifested itself by a *"conical light beam"*, This effect is caused by the colloidal particles which diffract light at all possible angles Θ. Faraday observed the same effect in gold salts. When the size of the colloidal particles of gold increases, the colour of the scattered light changes from red to blue.

A typical example of the Tyndall effect in glasses is the light scatter caused by colloidal particles (crystals) of gold in gold ruby glass. In gold ruby glass the colour of the scattered light also changes, if these particles grow larger.

A beautiful purple colour in gold ruby glass is obtained if the size of the colloidal particles ranges between 20 nm and 50 nm. In that case, no scattering occurs. If, however, the particles reach a size of $70-100$ nm, a slight scatter can be observed, which becomes more pronounced when the particles reach a size of $200-500$ nm. In this latter case, the metallic lustre of the particles of gold will scatter the light. Other colloidal particles of metals, too, e.g., of silver, copper, tin, lead, platinum, rhodium, paladium etc., can give rise to the Tyndall effect, if they have suitable dimensions.

Another example is light scatter in opals. Particles that are present in the form of crystallized droplets of $Ca_3(PO_4)_2$ scatter light without colouring it visibly if their size is less than the wavelength of the incident light. When the size increases, the scattered light has a bluish tint of increasing intensity, despite the fact that the transmitted light has a yellow-to-brownish colour.

The effect of the light scatter, which was first observed by Tyndall, has since been theoretically formulated and interpreted by Rayleigh, Mie, Gans and others.

186

Rayleigh scatter

According to Rayleigh's definition [2, 3], the scattering of light by centres that are small with respect to the wavelength of the light, λ $\left(\text{e.g.} < \dfrac{\lambda}{4\pi}\right)$ must in principle be viewed as a *diffraction phenomenon*. If the electromagnetic radiation meets a scattering centre, each electron becomes the source of secondary light, the frequency of which is, to a large extent, identical with the frequency of the primary light (approximately 0.1 % being of a different frequency). The light is emitted in all directions and the intensity of this scattered light depends upon the angle of the measurement. If the scattering centres are isotropic, the scattered light, measured at the angle of 90°, will always be fully polarized in the plane normal to the direction of the incident light, whether the primary light has been polarized or not. If the scattering centres differing in their indices of refraction from the parent phase are not isotropic, partial depolarization will occur.

Using natural, unpolarized incident light, Rayleigh derived the relationship.

$$\frac{\Phi_{\tau d}}{\Phi} = \frac{9\pi^2 N}{2d^2\lambda^4}\left(V\,\frac{n^2 - 1}{n^2 + 2}\right)(1 + \cos^2\Theta), \tag{223}$$

where $\Phi_{\tau d}$ is the intensity of the scattered radiant flux,
     Φ the intensity of the incident radiant flux.
     d the distance of the observed detector from the scattering centre,
     λ the wavelength of the incident radiant flux,
     N the number of the scattering centres in a specific volume,
     n the ratio of the refractive index of a scattering centre to the refractive index of the solvent,
     V the volume of the scattering centre,
and Θ the angle of observation.

Figure 132 shows $\Phi_{\tau d}$ as a function of the angle of scatter in optical glass for a wavelength of 546 nm (b).

The area enclosed by the ends of the vectors emerging from an infinitely small element of the surface, or volume, of the scattering body (the vectors standing either for the relative luminous intensity or for the relative luminance of the element in the direction considered) is termed the *indicatrix of scatter*. It is usually represented with the help of polar coordinates, but it can also be represented by means of rectangular coordinates. In order to simplify the interpretation of the results, we usually employ a *vertically polarized incident radiant flux*, Φ, so that the intensity of Rayleigh scatter is independent of the angle of measurement, Θ (see Figure 132 b). By vertically polarized light we mean light polarized perpendicularly to the plane of the measured intensity of the scattered light.

The Rayleigh relationship (223) holds only for low concentrations. In highly concentrated systems, independent scattering centres can hardly be considered, and the intensity of the scattered radiant flux recorded will thus not be proportional to the concentration.

In glasses, which from a structural point of view are looked upon as amorphous materials, light scatter should theoretically not be recorded. A weak Rayleigh scatter, which we do record, e.g., in the relatively homogeneous optical glasses (see Figure 132 b), is, according to Prod'homme [3] and others, caused by *thermal fluctuations of the density*, and is thus proportional to absolute temperature *T*.

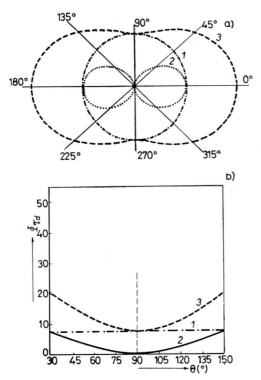

Fig. 132 The polar diagram (a) and the indicatrix curve (b) for Rayleigh scatter in optical glass: *1* — incident light vertically polarized; *2* — incident light horizontally polarized; *3* — incident light unpolarized.

In view of the fact that the density fluctuations depend upon the isothermal compressibility, χ, Einstein suggested that the intensity of the scattered radiant flux may be written

$$\frac{\Phi_{\tau d}}{\Phi} = \frac{\pi^2 V^2}{2d^2 \lambda^4} (n^2 - 1)^2 \, kT\chi, \tag{224}$$

where **k** is the Boltzmann constant,
and   *T* is the absolute temperature.

But Maurer [4] reasons that if the scattering centres are small compared with the wavelength of the radiation, glasses do not virtually differ from liquids, and the intensity of the scattered radiant flux should, in his opinion, largely depend

188

upon *fictive temperatute*, τ. A major part of the scattered radiant flux can be ascribed to the fluctuations of the concentrations present in the glass after the latter has been cooled from the liquid state. And Volkenštejn [5] shares this view.

It has long been assumed that if the scattering centres are substantially smaller than the wavelength of the incident radiation, the intensity of the scattered radiant flux should, according to equation (223), be inversely proportional to the fourth power of the wavelength, λ^4. This relationship is valid for gases only, it does not hold for crystals or liquids.

Prod'homme [6] has discovered that in glasses the exponent in equation (223) will range between the values 4 and 8, according to the composition of the glass.

Concluding this section, we can thus state that if the Rayleigh condition $\left(\text{viz. the scattering centres smaller than } \dfrac{\lambda}{4\pi}\right)$ is satisfied, the shape of the indicatrix for a vertically polarized incident radiant flux will invariably be circular. From a number of measurements carried out by various authors it follows that nearly all glasses in which scattering centres have deliberately not been established will manifest this type of scatter.

Mie scatter

As soon as the dimensions of the scattering centres have reached a value corresponding to the wavelength, λ, of the incident radiant flux, the penetration of the scattering centre will take some time, which is, of course, by no means negligible as compared with the period of the electromagnetic field. It becomes necessary therefore to introduce a *factor expressing the phase shift related to the angle of scatter*.

Mie [7, 3] investigated this problem in detail and has formulated a mathematical theory of the scatter of light by larger spherical scattering centres suspended in the solvent. In his computations he considered the *electric and magnetic multipole moments* induced by the electromagnetic field in the scattering centres.

If $k = 2\pi/\lambda$, we can write for the given moments

$$a_1 = 2k^2R^3 \frac{n^2 - 1}{n^2 + 2}, \tag{225}$$

$$a_2 = \frac{k^5R^5}{6} \frac{n^2 - 1}{n^2 + 1.5}, \tag{226}$$

$$p_1 = -\frac{k^5R^5}{15}(n^2 - 1), \tag{227}$$

where a_1, a_2 are the dipole and quadrupole electric moments,
      p_1 is the dipole magnetic moment,
and R is the radius of the scattering centre.

189

For the vertically polarized component the intensity of the scattered radiant flux can then be expressed by the relationship

$$\frac{\Phi_{\tau d}}{\Phi} = \frac{N\lambda^2}{8\pi^2 d^2}\left(\frac{a_1}{2} + \frac{a_2 + p_1}{2}\cos\Theta\right)^2,$$

(228)

and for the horizontally polarized component by the relationship

$$\frac{\Phi_{\tau d}}{\Phi} = \frac{N\lambda^2}{8\pi^2 d^2}\left(\frac{a_1\cos\Theta}{2} + \frac{a_2\cos 2\Theta}{2} + p_1\right)^2.$$

(229)

For vertically polarized light the indicatrices are no longer circular (symmetrical), as was the case with Rayleigh scatter, but they exhibit higher values of the intensity of the scattered radiant flux for lower angles, and vice versa, often with several maxima and minima.

Figure 133 gives both the polar diagram (a) and the scattering curve (b) for glass containing colloidal palladium and silver at 546 nm.

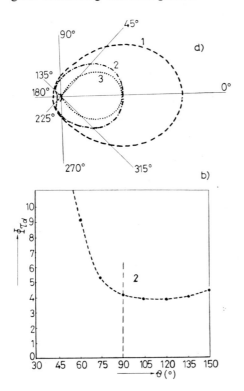

Fig. 133 The polar diagram (a) and the indicatrix curve (b) for Mie scatter in photosensitive glass containing Pd and Ag: 1 — incident light is unpolarized; 2 — incident light vertically polarized; 3 — incident light horizontally polarized.

This type of scattering characteristic can be employed to identify various inhomogeneities in glass manifesting a unique change in the local polarizability, e.g., bubles, impurities, suspensions, colloidal particles of metals, inclusions, etc. The inherent structural heterogeneities which cause polarizability to fluctuate over

190

a radial distance r (e.g., phase separation etc.) cannot, however, be fully accounted for by this theory.

Rayleigh-Gans scatter

Gans [8] has shown that if the refractive index of the scattering centre differs from the refractive index of the surrounding medium by ± 0.05, the effects influencing the electromagnetic field of the large scattering centres present will be substantially restrained, and the whole problem will be simplified a great deal compared with the Mie theory.

The Rayleigh-Gans approximation theory of scatter postulates that each volume element of a scattering centre, V, acts independently of the other elements as in Rayleigh scatter. However, depending on how the volume elements of the scattering centres are spatially located, the scattered light waves will manifest a mutual *phase difference*, δ, giving rise to *interference effects*.

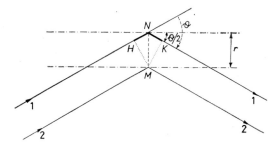

Fig. 134 Phase difference δ of light between points M and N at distance r.

Figure 134 is a diagram of the phase difference, δ, of a light wave incident on two points, M and N, separated by a distance r, for which we can write

$$\delta = 4\pi r \sin\left(\frac{\Theta}{2}\right)\Big/\lambda. \tag{230}$$

Compared with the ray that is diffracted at point M, the path of the ray diffracted at point N will thus be longer by $HN + NK = 2MN \sin(\Theta/2)$.

For natural incident light the value of the intensity of the scattered radiant flux can be expressed as

$$\frac{\Phi_{\text{rd}}}{\Phi} = \frac{4\pi^2 V^2}{d^2 \lambda^4}(n_0 - 1)^2\, \Phi(\Theta, \lambda)(1 + \cos^2 \Theta), \tag{231}$$

where n_0 is the mean value of the refractive index
and Φ is the scattering function.

For the sake of simplicity, this theory postulates a scattering centre the mean value of the variable refractive index of which, n_0, coincides with the index of refraction of the glass environment. This scattering centre has a nucleus, whose refractive index $n_1 > n_0$, and an enveloping concentric layer with the refractive

191

index $n_2 < n_0$, so that the distribution of the index of refraction is as shown in Figure 135 a.

However the scattering centre has actually no strictly defined phase boundary, for the interface region is an ionic transfer medium, whether we consider phase separation, crystallization, or a heterogeneity in the medium. The change in the index of refraction is thus not discontinuous, as was the case with the model centre (Figure 135 a), but continuous, as schematically shown in Figure 135 b.

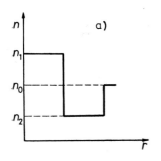
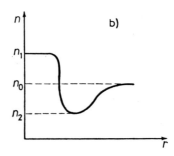

Fig. 135 Refractive index in a model scattering centre with concentric layers (a), and in a real scattering centre (b).

Figure 136 presents the scattering characteristic (*a*) and the results of the measurement of the value of the intensity of scattered radiant flux Φ_{rd} as function of Θ, in borosilicate glass for $\lambda = 546$ nm (*b*). The glass was subjected to thermal treatment at various temperatures $t_3 > t_2 > t_1$. The reader will see that the indicatrix is uniformly extended in the direction towards the larger angles, where the peak value of scattered radiation was recorded.

A typical field of application of the Rayleigh-Gans theory is the scattering of light by separated phases in glass.

Mandelshtam-Brillouin scatter

So far we have assumed that the spectral distributions of the scattered light and the incident light are identical. Studying the problem in more detail Mandelshtam, Brillouin [9, 10] and Schroeder [39] have however found theoretically, and Gross has experimentally proved, that in crystals the frequency of incident radiation gives rise to six frequencies; in amorphous materials (glasses) the distribution is into four frequencies, and in liquids into two only. Figure 137 (after [10]) presents the results of the measurement of the Brillouin scatter in a quartz crystal (*a*), quartz glass (*b*), and Pyrex glass (*c*).

This theory is based upon the assumption that light may be scattered by acoustic waves of high (ultrasonic) frequency. These waves are associated with the phonons in the material and hence may be related to concentration fluctuations.

Raman scatter

Apart from the acoustic vibrations observed by Mandelshtam and Brillouin, optical vibrations must also be taken account of. The light scattered by a crystal must thus be modulated by both the acoustic and the optical vibrations. Modulation of this kind gives rise to new lines in the spectrum of the scattered light, the frequency of which differs from the frequency of the incident radiation. This effect has been observed, e.g., in quartz crystal, but also in glasses, and is termed the

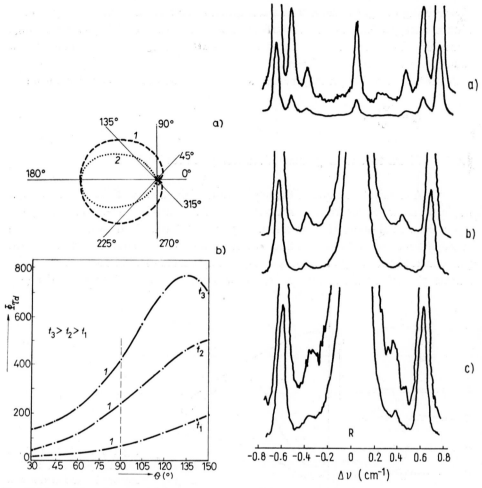

Fig. 136 Intensities of scattered light in the Rayleigh-Gans approximation (a), and the indicatrix curves (b) for thermally treated boro-silicate glass: *1* — incident light vertically polarized; *2* — incident light horizontally polarized.

Fig. 137 Brillouin spectra of quartz crystal (a), quartz glass (b), and Pyrex glass (c); R = Rayleigh line.

Raman scatter. Raman scatter involves a combination of the vibrations of the light wave and the internal vibrations of the molecules. In contrast to the classical Rayleigh scatter, the Raman scatter is incoherent; the phases are randomly distributed, owing to the fact that each molecule vibrates independently of the others. On either side of the Rayleigh line thus lies the Raman spectrum affected by both the vibrations and the rotation of the molecules.

Definition of scatter

Scattering of light is defined as a change in the direction of the propagation of radiant energy arising from an interface between two media, or in passing through a medium. Uniform scatter is the ideal, the scatter being such that luminance, or radiance, are equal, in all directions.

Scattering of radiation may be defined with the help of the so-called turbidity.

$$\tau_d = \frac{\Phi_{\tau d}}{\Phi}. \tag{232}$$

4.4.3 Character and magnitude of scattering centres in glasses

A classical example of the application of the theory of the scatter of light in glasses is the scatter from separated phases. The scattering indicatrix will depend upon the actual state of separation. Figure 138 (after Prod'homme [3]) is a schematic representation of the fluctuations of the refractive index in the separated phase at different stages and the corresponding curves of the indicatrix for soda-silicate glass. (In the measurements, use was made of vertically polarized light.)

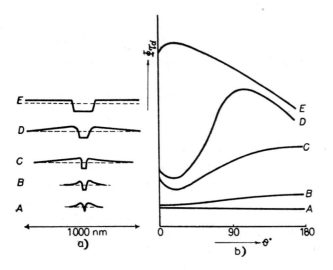

Fig. 138 Fluctuation of the refractive index in the separated phase at different stages of separation (a), and the corresponding indicatrix curves (b) for soda-silicate glass.

If n_0 is the refractive index of the parent glass, and if we assume the separated phase to be enriched by silicon dioxide, the index of refraction of that phase n_1 will be lower than that of the parent glass n_0, $n_1 < n_0$. The edge zone will manifest a refractive index higher than the parent glass, $n_2 > n_0$. In Figure 138, A indicates

194

the onset of the separation process. The scatter is very poor and the angular distribution is symmetrical, so that the Rayleigh condition of scatter is satisfied.

B corresponds to a scattering centre of order 20 nm. The ege field is a quite a distance, thus the indicatrix is extended in the direction towards larger angles.

C is a diagram of the nucleus the diameter of which is about 40 nm; the dimension of the surrounding peripheral field is several hundreds of nanometers. The indicatrix has a marked extension towards the larger angles.

D refers to a nucleus the diameter of which is about 100 nm, the surrounding field undergoes disintegration and combines with the field of the adjoining separated phases. The scattered light is very intensive. For *B* and *C* the curves of the indicatrices can thus be interpreted in terms of the Rayleigh-Gans approximation; *D* is a transitional type.

In *E* the diffusion region unites again, in the glass there no longer exists any medium with refractive index n_0, the separation of the phases is completed, and the scattering centres with the refractive index n_1 are now located in the n_2 refractive index medium. The scattered light has its peak at smaller angles and the indicatrix curves can be characterized by Mie scatter. We assume that the diffusion region has reached a minimum and the separated scattering centres are sharply defined.

The *first perceptible scatter in glasses* can be seen visually when the scattering centres reach a diameter of $10-20$ nm. If this dimension exceeds about 100 nm, and the concentration of these centres in a specific volume is high, scatter will be so great that such glasses will appear to be cloudy. If the scattering centres reach a diameter of approximately 1000 nm diameter, the glasses will appear opaque.

If the scattering centres satisfy the conditions of the Rayleigh-Gans approximation (Figure 139, curve *L*), the indicatrix curves should theoretically increase monotonically towards larger angles [11]. But this basic effect is actually very often interfered with by the scattering centres satisfying the Mie scatter (Figure 139, curve *S*) and causing an increase in the intensity of the scatter at smaller angles. The *superposition of the two types of the scattering centres* then gives an indicatrix which exhibits a wide minimum in the region 45°, as seen in Figure 139.

If the indicatrix curve has a maximum, the dimensions of the scattering centres can be calculated according to Prod'homme [11]. This theory is probably based on Mandelshtam's assumption, viz. that the scatter of X-ray radiation by a crystal lattice is similar to the scatter of light from fluctuations of concentrations, which are periodically recurrent. Given the conditions quoted in [11], the actual diameter of the scattering phase can be computed from the relationship

$$D = \sqrt{\frac{5}{3}} \frac{\lambda}{4 \sin \Theta_m/2},$$ (233)

where Θ_m is the scattering angle corresponding to the indicatrix maximum.

Prod'homme [11] has also suggested a graphical method of computing the dimensions of the scattering centres of separated phases. He devised a curve (see Figure 140) of the dependence of the dissymmetry $z = \dfrac{\Phi_{\tau d}\,45^0}{\Phi_{\tau d}\,135^0}$ upon $\dfrac{\pi D}{\lambda}$, where D is the dimension of the scattering centre. He assumed that if the dimension of the scattering centre reaches low values and the scattering centres are of the Rayleigh type, the dissymmetry value z approaches a limiting value of 1. The curve has its minimum at $\dfrac{\pi D}{\lambda} = 0.5$ (calculated after Goldstein [12]); with increasing values of $\dfrac{\pi D}{\lambda}$, the values of z will rise too. The dimensions of the centres are already so large in this region that maxima may form on the curves, from which the dimensions can be computed with the help of equation (233).

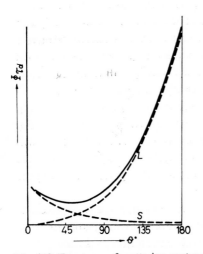

Fig. 139 Two types of scattering centres superimposed:
S — Mie-type scattering centres L — Rayleigh-Gans scattering centres.

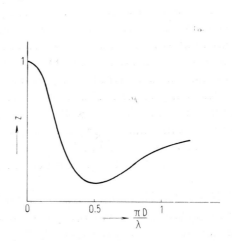

Fig. 140 Variation of dissymmetry $z = \dfrac{\Phi_{\tau d45^0}}{\Phi_{\tau d135^0}}$ with $\dfrac{\pi d}{\lambda}$.

In the literature we often come across the terms *dissymmetry, depolarization,* and the *Rayleigh ratio.* In their simplest definitions these terms refer to the scattering characteristic of glasses, and we shall therefore briefly deal with their meanings.

Dissymmetry z

Dissymmetry is usually defined by the ratio of the intensity of the radiant flux at 45°, $\Phi_{\tau d\,45^\circ}$, to the intensity of the radiant flux scattered at 135°, scattered $\Phi_{\tau d\,135^\circ}$, thus

$$z = \frac{\Phi_{\tau d\,45^\circ}}{\Phi_{\tau d135^\circ}}. \qquad (234)$$

196

If the scattering centres satisfy the Rayleigh condition, viz. if they are smaller than $\dfrac{\lambda}{4\pi}$, then for the vertically polarized incident light it follows that $z = 1$.

The indicatrix is thus symmetrical round the scattering centre (see Figure 132). If the scattering centres can be characterized by Mie scatter, or, on the contrary, by the Rayleigh-Gans approximation, the numerical values of z will be greater than, or less than, unity. In some cases use is made of the diffusion factor (CIE standard), which gives the ratio of the arithmetic mean of the values of luminance measured at angles of 20° and 70° and of luminance at an angle of 5° from the normal for perpendicular incidence of light [13].

Depolarization ϱ'

Depolarization is defined as the ratio of the intensity of the horizontally polarized component to the intensity of the vertically polarized component of the scattered light. H denotes the intensity of the scattered, horizontally polarized component, V the intensity of the vertically polarized component. According to whether the incident light has been polarized horizontally or vertically, subscripts h or v are added; the unpolarized light has subscript n. Depolarization can thus be expressed by the relationship

$$\varrho_i' = \frac{H_i}{V_i}, \qquad\qquad (235)$$

where n, v, h are substituted for i.

Using equation (235) we can thus calculate one of the quantities of ϱ_i' if the other two are known.

The Rayleigh ratio

The Rayleigh ratio is defined by the relationship

$$Re\,90^0 = \frac{\Phi_{\tau d}\,90^0 d^2}{\Phi V}, \qquad\qquad (236)$$

where $Re\,90°$ is the Rayleigh ratio for the angle $\Theta = 90°$,

$\Phi_{\tau d}\,90°$ the intensity of the scattered radiant flux at $\Theta = 90°$.
   Φ the intensity of the incident radiant flux,
   d the distance of the observer from the scattering centre,
and V the scattering volume.

This ratio is made use of for the quantitative measurement of the intensity of the light scattered at 90°.

We further have the so-called *internal interference*, which is defined as interference within the same scattering centre, and *external interference*, defined as interference in different, spatially related scattering centres. These characteristics are mostly utilized in the evaluation of the dimension, concentration, and character of the scattering centres in solutions.

4.4.4 Dependence of light scattering upon the chemical composition and physical state of glasses

In this section we shall present some results of the measurement of the intensity of scattered light in glasses of several types.

Scattering of light in thermally treated glasses of the
$Na_2O - B_2O_3 - SiO_2$ system

Prod'homme [11] studied the scattering characteristics of four borosilicate glasses subjected to different thermal treatment. The compositions of these glasses are given in Table 35.

Table 35

Glass	Mole %		
	SiO$_2$	B$_2$O$_3$	Na$_2$O
A	78.8	11.3	9.9
B	72.8	19.4	7.8
C	60.0	30.0	10.0
D	68.0	27.0	5.0

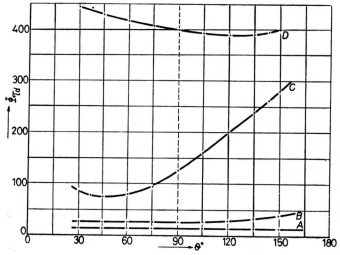

Fig. 141 Dependence of the intensity of scattered luminous flux Φ_{rd} upon the angle of measurement, Θ, in Na$_2$O—B$_2$O$_3$—SiO$_2$ glass not subjected to thermal treatment.

The glasses were thermally treated at temperatures ranging from 640 °C to 740 °C for 0.5−6 hours, and the intensity of scattered light, Φ_{rd}, was measured in the angular range $\Theta \sim 30° - 150°$. Figure 141 gives the indicatrix curves of thermally untreated glasses for incident light of wavelength 546 nm (vertically polarized).

The shapes of the curves are clearly indicative of the three types of the scattering centres already mentioned. The indicatrix curve of glass *A* is an example of the presence of the Rayleigh type of scattering centres, and in conformity with Rayleigh's theory no dissymmetry of diffusion can be observed here.

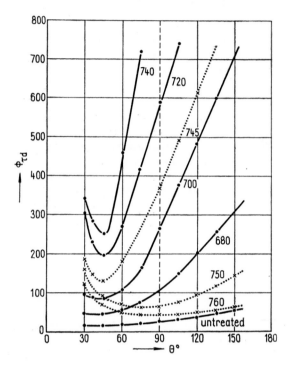

Fig. 142 Dependence of the intensity of scattered luminous flux $\Phi_{\tau d}$ upon the angle of measurement, Θ, in $Na_2O-B_2O_3-SiO_2$ glass. The incident luminous flux $\lambda = 546$ nm was vertically polarized. Glass *B* thermally treated for one hour at temperatures between 680—760 °C.

On the other hand, however, the indicatrix curves of glasses *B* and *C* correspond to scattering centres characterized by the Rayleigh-Gans approximation scatter, and the last indicatrix curve, the one of glass *D*, is characteristic of the scattering centres described by the Mie theory. Figures 142 and 143 give the scattering characteristics of glasses *B* and *C* thermally treated at various temperatures.

Scattering of light in thermally treated glasses of the $CaO-Al_2O_3-B_2O_3-SiO_2$ and $Na_2O-CaO-SiO_2$ system

Using measurements of light scatter, Hammel and Ohlberg [14] examined phase separation in thermally treated glass of the following composition (wt. %): 22 CaO, 14 Al_2O_3, 10 B_2O_3, and 54 SiO_2. Measurements were made for a wavelength of 436 nm. The results are presented in Figures 144, 145, and 146. In these Figures,

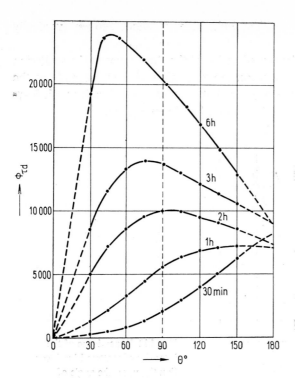

Fig. 143 Dependence of the intensity of scattered luminous flux $\Phi_{\tau d}$ upon the angle of measurement, Θ, in Na_2O—B_2O_3—SiO_2 glass. The incident luminous flux $\lambda = 546\,nm$ was vertically polarized. Glass C thermally treated for 0.5—6.0 hours at 725 °C.

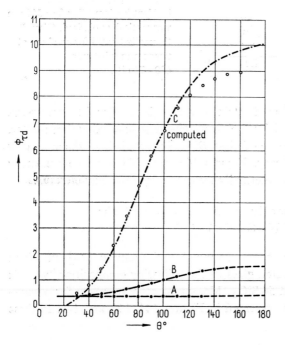

Fig. 144 Dependence of the intensity of scattered luminous flux $\Phi_{\tau d}$ upon the angle of measurement, Θ, in CaO—Al_2O_3—B_2O_3—SiO_2 glass. The incident luminous flux $\lambda = 436\,nm$ was vertically polarized. The computed theoretical curve for glass C is compared with experiment.

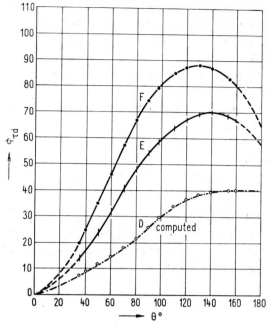

Fig. 145 Dependence of the intensity of scattered luminous flux $\Phi_{\tau d}$ upon the angle of measurement, Θ, in $CaO-Al_2O_3-B_2O_3-SiO_2$ glass. The incident luminous flux $\lambda = 436$ nm was vertically polarized. The computed theoretical curve for glass D is compared with experiment.

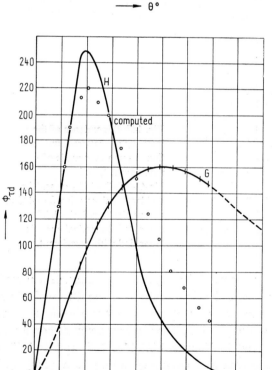

Fig. 146 Dependence of the intensity of scattered luminous flux $\Phi_{\tau d}$ upon the angle of measurement, Θ, in $CaO-Al_2O_3-B_2O_3-SiO_2$ glass. The incident luminous flux $\lambda = 436$ nm was vertically polarized. The computed theoretical curve for glass H is compared with experiment.

201

A	denotes the specimen of thermally untreated glass,
B	thermal treatment of the glass, 15 min at 801 °C (nucleation),
C	thermal treatment of the glass, 15 min at 801 °C (nucleation) and 8 min at 867 °C (growth),
D	thermal treatment of the glass, 15 min at 801 °C (nucleation) and 15 min at 867 °C (growth),
E	thermal treatment of the glass, 15 min at 801 °C (nucleation) and 20 min at 867 °C (growth),
F	thermal treatment of the glass, 15 min at 801 °C (nucleation) and 30 min at 867 °C (growth),
G	thermal treatment of the glass, 16 min at 840 °C,
and H	thermal treatment of the glass, 150 min at 840 °C.

The same method was used by Ohlberg and Hammel [15] in the examination of phase separation in a glass of the following composition (wt. %): 13 Na_2O, 11 CaO, 76 SiO_2. The glass subjected to a temperature 600 °C for 240 min exhibited the scattering characteristics presented in Figure 147.

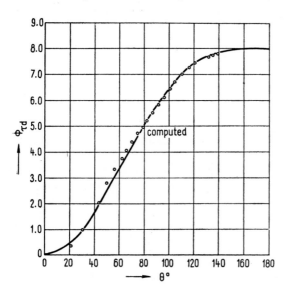

Fig. 147 Dependence of the intensity of scattered luminous flux Φ_{rd} upon the angle of measurement, Θ, in Na_2O—CaO—SiO_2 glass. The incident luminous flux $\lambda = 436$nm was vertically polarized. The computed theoretical curve is compared with experiment.

Ohlberg and Hammel [14] gave the curves of the angular distribution of scattered light a mathematical form. They based their procedure on the Rayleigh-Gans theory of scattering centres with spherical symmetry and they made calculations for the vertical component of the scattered light. The indicatrix curves computed and the values recorded are presented in Figures 144, 145, 146, and 147. The shapes of the indicatrix curves measured and computed testify to a very good agreement the theoretical interpretation and the results of the measurement. For a more complex case of calculation the reader is referred to reference [14].

Scattering of light in thermally treated photochromic glasses

Fanderlik and Prod'homme examined a glass containing a separated phase of silver chloride [16]. Figures 148 and 149 show the indicatrix curves of thermally treated photochromic glasses recorded for vertically polarized light of the wavelength 546 nm.

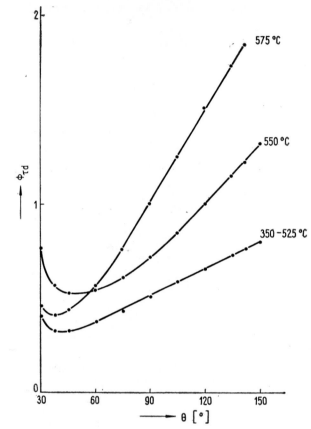

575 °C

550 °C

350 – 525 °C

Fig. 148 Dependence of the intensity of scattered luminous flux Φ_{ed} upon the angle of measurement, Θ, in photochromic glass. The incident luminous flux $\lambda = 546$ nm was vertically polarized. Glasses subjected to thermal treatment at 350—575 °C for one hour.

A one-hour thermal treatment at temperatures ranging from 350 °C to 525 °C does not affect the shape of the curves very much. The glasses manifest weak scatter and two types of scattering centres are present. The Mie scattering centres increase the value of scattered radiation at low angles, whereas the scattering centres satisfying the Rayleigh-Gans approximation increase the intensity of scattered light at higher angles. The indicatrix curves exhibit a wide minimum at the angle of 45° (see Figure 148).

At an advanced stage of separation the Mie centres will gradually disappear at temperatures as low as 600 °C, and the indicatrix will correspond to Rayleigh-Gans scatter. With increasing temperature of thermal treatment, and increasing phase

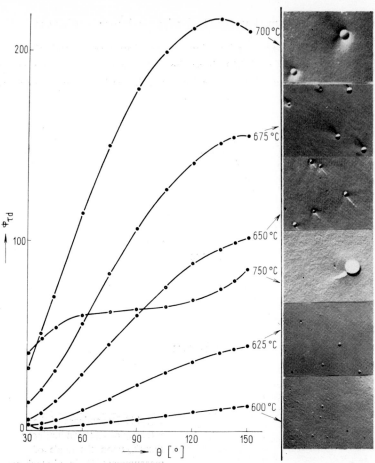

Fig. 149 Dependence of the intensity of scattered luminous flux $\Phi_{\tau d}$ upon the angle of measurement, Θ, in photochromic glass. The incident luminous flux $\lambda = 546$ nm was vertically polarized. Glasses were subjected to thermal treatment at 600°—750 °C for one hour.

Table 36

t [°C]	Size [nm]	t [°C]	Size [nm]
550	26	650	68
575	31	700	100
600	36	750	150
625	50		

separation, the indicatrix curves exhibit a continually increasing intensity of light towards larger angles. Table 36 lists the dimensions of the separated phase for various temperatures of the thermal treatment. The calculation was made with the help of the method discussed in Section 4.4.3.

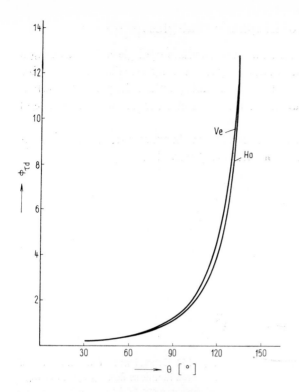

Fig. 150 Dependence of the intensity of scattered luminous flux $\Phi_{\tau d}$ upon the angle of measurement, Θ, in fluorophosphoric opal glass. The incident luminous flux $\lambda = 546$ nm vertically (Ve) and horizontally (Ho) polarized.

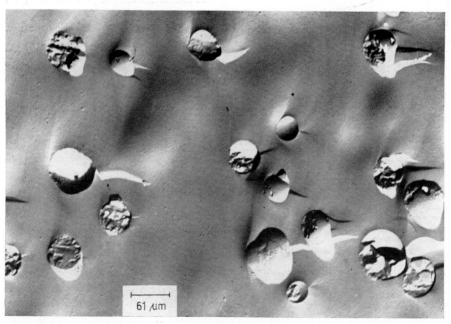

Fig. 151 Electron micrograph of fluorophosphoric opal glass.

205

Scattering of light in fluorophosphate opal glass

Exnar [37] measured light scatter in fluorophosphate opal glass for wavelength 546 nm and for vertically and horizontally polarized lights. The results of his measurement are presented in Figure 150, and a photograph of the particles (NaF + + CaF_2 + amorphous phase) taken on an electron microscope can be found in Figure 151.

A theoretical interpretation of the recorded indicatrix curve poses great problems owing to the high concentration of the scattering particles.

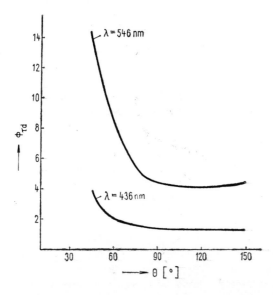

Fig. 152 Dependence of the intensity of scattered luminous flux Φ_{rd} upon the angle of measurement, Θ, in photosensitive glass containing Pd and Ag. The incident luminous flux $\lambda = 546$ nm and $\lambda = 436$ nm was vertically polarized. Glass not subjected to thermal treatment.

Scattering of light in photosensitive glass containing palladium and silver

An indicatrix curve of a shape typical of the colloidal particles of metals in glass is presented in Figure 152 for photosensitive glass containing palladium and silver. The wavelength of the light (vertically polarized) equalled 546 nm and 436 nm.

The shapes of the indicatrix curves are clearly indicative of Mie scatter.

4.4.5   Measuring methods and apparatus

The intensity of the scattered radiant flux, Φ_{rd}, which is dependent upon the angle of measurement, Θ, is generally measured with the help of *goniophotometers*. The simpler types facilitate the measurement of the intensity of the light scattered at angle of 90°.

206

Some of these instruments are combinations of the standard optical units, others are stock types produced commercially. We shall now deal with the commercially produced types of instrument.

The ARL-FICA goniophotometer

ARL-FICA, a French firm, is the producer of the P.G.D 42 000 M goniometer, a schematic diagram of which is shown in Figure 153.

As a prerequisite for the measurement of the intensity of scattered light in its dependence upon the angle of measurement, a cylinder of 15 mm base diameter and 30 mm height must be cut from the glass and its surface optically polished.

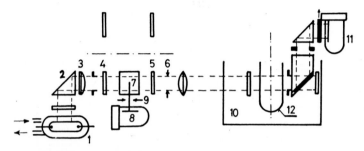

Fig. 153 Schematic optical diagram of the ARL-FICA 42000 goniophotometer: *1* — high pressure mercury vapour lamp; *2* — prism; *3* — condenser; *4* — carrousel with filters; *5* — polarizer; *6* — iris diaphragm; *7* — reference diffusion element; *8* — photocell; *9* — iris diaphragm; *10* — vessel containing toluene or benzene with immersed specimen to be measured; *11* — photomultiplier, type RCS IP 28, with an optical system, capable of swivelling the specimen through the angle 0°—180°; *12* — specimen of the glass to be measured.

The glass specimen is then placed in a tank filled with an immersion liquid (see Figure 153) and the relative value of the intensity of scattered radiant flux, Φ_{rd}, recorded within the range of 30°—150°.

The measurement must be carried out with the intensity of the incident radiant flux, Φ, constant, and the values of the relative intensity of scattered light are multiplied by sin Θ, so that the recorded values may be related to the specific volume of the glass.

SPEKOL spectral colorimeter with FK adapter

One of the instruments used for measuring the intensity of the light scattered at 90° is the SPEKOL spectral colorimeter with the FK adapter and amplifier. The apparatus operates on the principle of the single-beam reflection method. The grating monochromator permits the separation of the individual spectrum lines if a mercury vapour lamp is used as the light source. An optical schematic diagram of the FK adapter is shown in Figure 154.

The ray of light passes through the slit (*1*) on to the lens (*2*), and continues through the diaphragm (*3*) to meet the glass specimen (*4*). With the help of an optical system of lenses (5, 7), or a filter (6), part of the light scattered at 90° is focussed on to the gas-filled phototube EGS (9), which is sensitive to radiation of the 360 nm – 630 nm wavelengths. The photocurrent is amplified by an amplifier capable of changing sensitivity over a wide range. The in-built measuring device then gives the values of the intensity of the scattered radiant flux.

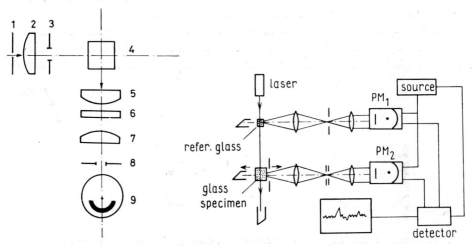

Fig. 154 Schematic optical diagram of the FK adapter for the measurement of luminous flux scattered at 90°.

Fig. 155 Schematic diagram of the equipment for the measurement of the homogeneity of glasses by the method of scattered luminous flux.

The accuracy of the methods mentioned, and particularly the interpretation of the results of the measurement, depend upon the homogeneity of the glass specimens, because the non-homogeneities, which are not the object of the examination, can, with their own characteristics superimposed upon the scattering characteristics of the heterogeneities to be examined, bias the results of the measurement.

Apparatus for measuring light scattering spectra

After Schroeder [39], the experimental techniques for measuring light scattering spectra of glasses are restricted to two basic systems:

(a) a dispersive method to obtain Rayleigh and Brillouin spectra using some type of high-resolution optical Fabry-Perot spectrometer as an optical filter,

(b) a system employed mostly for obtaining correlation functions and thus probing the Rayleigh linewidths; a type of digital correlation spectrometer.

Use of light scattering in inhomogeneity tests

Prod'homme [17] has used light scattering for the examination of micro-heterogeneous centres in optical glasses. A schematic diagram of the arrangement of the measuring equipment is shown in Figure 155.

The source of light used was a He — Ne gas laser. Having passed through the reference diffusion element, the $\lambda = 633$ wavelength radiation meets the measured specimen of glass, which can be moved in the direction perpendicular to the incident ray of light with the help of a micrometer screw.

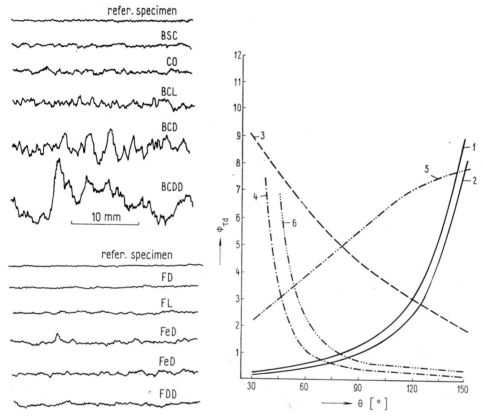

Fig. 156 Results of the measurement of the homogeneity of optical glasses.

Fig. 157 Dependence of the intensity of scattered luminous flux $\Phi_{\tau d}$ upon the angle of measurement, Θ, for:
1 — Czechoslovak Švedlár quartz vein material,
2 — Czechoslovak Dětkovice quartz vein material;
3 — Soviet Kyshtym quartz material; 4 — Brazilan crystal; 5 — Soviet MBK-Svetlorechensk quartz material; 6 — the US Arkansas I quartz material.

At an angle of 90° to the incident light the values of the intensity of scattered light are related to the shift of the glass specimen. The results of the measurement for several types of optical glasses are presented in Figure 156.

Scattering of light in natural quartz materials

Fanderlik and Dubský [38] used the ARL-FICA 42 000 goniophotometer to measure the intensity of scattered radiant flux $\Phi_{\tau d}$ from natural quartz materials of various geological types. Owing to the different conditions of their formation, these raw materials vary in the type and the number of their scattering centres due to pressure-induced disturbances, different types of liquid, fluid, and gaseous inclusions, inclusions of other minerals, the type and number of the structural defects, etc.

The measurement was carried out with vertically polarized flux Φ at a wavelength of 436 nm.

The silica material ($n_{436\,nm} = 1.547$ to 1.549) of particle size 0.1 mm to 0.5 mm was contained in a cylindrical cell together with the immersion liquid, viz benzyl alcohol p.a. ($n_{436\,nm} = 1.549$, at 20 °C). For a description of the measuring procedure the reader is referred to the previous paragraphs. The results of the measurement are shown in Figure 157.

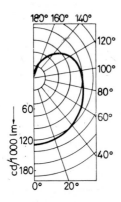

Fig. 158 Polar diagram of a lighting fixture.

In spite of the fact that the following paragraphs may seem to be out of context, we shall briefly deal with the questions of *measuring the luminous intensity of a light source*, or a lighting unit, because opal glasses are extensively used in the production of lighting equipment.

Light sources and lighting fixtures are characterized by their *curves of luminous intensity*. Since most lighting fittings are rotationally symmetrical, the evaluation of luminous intensity can be made with the sole help of the photometric surface section, the so-called luminous intensity curve. The latter can be plotted from the data obtained in such a way that a plane running through the axis of the lighting fixture is selected and the luminous intensity of the source, or the fitting, measured in different directions (angles). The values recorded are then used for plotting a graph of the respective fixture (see Figure 158, after [33, 40]).

In evaluating various lighting equipment, use must be made of a *standard light source* (generally 1000 lm luminous flux), otherwise the values recorded must be converted into those corresponding to the luminous efficacy of that source.

210

In asymmetrical lighting fittings the luminous intensity curves plotted for a number of planes give the so-called *isocandela diagram*. Measurement is carried out with the help of the so-called *distribution photometer*.

Measurements are taken in an integrating sphere and the radiant flux emitted in all directions is recorded. As already mentioned above, the etalon of the radiant flux is ascertained first, and only then the source of light, or the lighting fitting, itself measured.

For more details the reader is referred to references [33, 40].

References

[1] J. Tyndall, *Phil. Mag.* On the blue colour of the sky, the polarisation of sky-ligh* and on the polarisation of light by cloudy matter generally. *37* (1869) p. 384.

[2] J. W. Rayleigh, On the electromagnetic theory of light. *Phil. Mag.*, *39* (1871) pp. 384–392.

[3] L. Prod'homme, Diffusion de la luniére et structure des verres. *Revue d'opt.*, *45* (1966) No. 4, pp. 163—176.

[4] R. D. Maurer, Light scattering by glasses. *J. Chem. Phys.*, *25* (1956) No. 6, pp. 1206—1209.

[5] M. V. Volkenštejn, Struktura a fyzikální vlastnosti molekul (Structure and Physical Properties of Molecules). NČAV, Prague 1962.

[6] L. Prod'homme, Répartition spectrale de la lumiére diffuseé par les verres. *Verres Réfract.*, *18* (1964) No. 1, pp. 10—18.

[7] G. Mie, Optik trüber Medien, speziell koloidaler Metallösungen. *Ann. Physik*, *25* (1908) pp. 377–381.

[8] R. Gans, Lichtzerstreung inforlage der molekularen Raubigkeit der Treunungsfläche zweier durchsichtiger Medien. *Ann. Physik*, *76* (1925) p. 29.

[9] L. Prod'homme, La diffusion Rayleigh et Brillouin dans les verres minéraux. *Verres Réfract.*, *28* (1974) p. 1, pp. 3—15.

[10] S. M. Shapiro, R. W. Gammon, and H. Z. Cummins, Brillouin scattering spectra of crystalline quartz, fused quartz and glass. *App. Phys. Lett.*, *9* (1966) No. 4, p. 157.

[11] L. Prod'homme, Diffusion de la lumiére par des verres et séparation de phases. *Verres Réfract.*, *22* (1968) No. 6, pp. 604—613.

[12] M. Goldstein, Theory of scattering for diffusion-controlled phase separations. *J. Appl. Phys.*, *34* (1963) No. 7, p. 1928.

[13] ČSN 36 0000 Světelné technické názvosloví (Lighting terminology). After IEC and CIE. 1967.

[14] J. J. Hammel and S. M. Ohlberg, Light scattering from diffusion-controlled phase separation in glass. *J. Appl. Phys.* *36* (1965) No. 4, pp. 1422—1447.

[15] S. M. Ohlberg, H. R. Golog, J. J. Hammel, and R. R. Lewchuk, Non-crystalline microphase separation in soda-lime-silica glass. *J. Am. Ceram. Soc.*, *48* (1965) No 6, pp. 331—332.

[16] I. Fanderlik and L. Prod'homme, Séparation de phases observeé par diffusion de la lumiére dans un verre photochrome. *Verres Réfract.*, *27* (1973) No. 3, pp. 97—104.

[17] L. Prod'homme, Analyse de l'homogénéité du verre par un faisceau laser. *Verres Réfract.*, *26* (1972) No. 6, pp. 167—174.

[18] H. C. Van de Hulst, Light Scattering by Small Particles. Wiley, New York 1967.

[19] R. D. Maurer, Crystal nucleation in a glass containing titania. *J. Appl. Phys.*, *33* (1962) No. 6, pp. 2132—2139.

[20] L. Prod'homme, Étude de la Démixtion de certains verres par mesure de leur transmission. *Verres Réfract.*, *25* (1971) No. 2, pp. 71—80.

[21] H. Rötger, Theoretische Auswertung von Trübungsuntersuchungen auf einem Fluor-Silikatglas. *Silikat Technik*, (1964), No. 3, p. 71.

[22] L. Prod'homme, Diffusometre visuel. *Rev. d'Optique*, *32* (1953) p. 615.

[23] A. Moriya, D. H. Warrington, and R. W. Douglas, A study of metastable liquid-liquid immiscibility in some binary and ternary alkali silicate glasses. *Phys. Chem. Glasses*, *8* (1967) No. 1, pp. 19—25.

[24] A. G. Koljagin, Anomalnoye raseyanie sveta v stekle (Anomalous scattering of light in glass). *Opt. i spektroskop.*, *6* (1956) No. 7, pp. 907—916.

[25] N. A. Voishillo, Coherent scattering of light in glass. *Opt. spektrosk.*, *12* (1962) pp. 412—418.

[26] L. Prod'homme and C. Vacherand, Étude par diffusion de la lumiére des variations structurales qui préparent la dévitrification. *Verres Réfract.*, *20* (1966) No. 5, pp. 354—361.

[27] B. Sedláček, Rozptyl světla — přehledný referát (Scattering of light — a survey report). *Chem. listy*, *47* (1953) No. 5, pp. 752—792; No. 6, pp. 908—944.

[28] M. Goldstein, Depolarised component of light scattered by glasses. *J. Appl. Phys.*, *30* (1959) pp. 493—500, 501—506; 33 (1962) pp. 3377—3382.

[29] R. S. Krishnan, Scattering of light in optical glasses. *Proc. Indian Acad. Sci.*, *3A*, (1936) p. 24.

[30] L. Mandelshtam, Polnoye sobraniye trudov (Collection of Works). Izdat. Akad. Nauk SSSR, Moscow, 1948.

[31] C. Tanford, Physical Chemistry of Macromolecules. Wiley New York 1961.

[32] V. E. Eskin, Rasseyanye sveta rastvorami polimerov (Scattering of Light in Polymeric Solutions). Izdat. Nauka, Moscow 1973.

[33] I. Fanderlik, Rozptyl světla sklem (Diffusion of Light by Glass). *Inform. přehled SVÚS Hradec Králové*, XVIII, (1975), No. 2.

[34] I. L. Fabelinkyi, Molekulyarnoye rasseyanye sveta (Molecular Scattering of Light). Nauka, Moscow 1965.

[35] T. Wu and T. Ohmura, Quantum Theory of Scattering. Prentice Hall, New York 1962.

[36] K. S. Shifrin, Rasseyanye sveta v mutnoy srede (Diffusion of Light in Turbid Medium). GITL, Moscow 1951.

[37] P. Exnar, Private communication, 1981.

[38] I. Fanderlik and F. Dubský, Úhlové rozdělení rozptýleného světelného toku nehomogennitami v křemenných surovinách (Angular distribution of scattered luminous flux due to non-homogeneities in quartz materials). *J. Geolog. Sci. Technology Geochemistry 17*, 1981, pp. 127—138.

[39] M. Tomozawa, R. H. Doremus, Treatise on materials science and technology, Vol. 12. Glass I.: Interaction with electromagnetic radiation. Light Scattering in Glass (J. Schroeder), Academic Press, New York, San Francisco 1977, pp. 157 to 222.

[40] Collective of authors, Osvětlovací sklo v interiéru (Interior Lighting Glass). SNTL, Prague 1965.

4.5 POLARIZATION, BIREFRINGENCE, AND INTERFERENCE OF RADIATION INTERACTING WITH GLASSES

4.5.1 Introduction

So far we have been concerned with the effects of reflection, refraction, absorption, and scattering in glass. Only occasionally has mention been made of the fact that the radiation may be polarized, viz. that its vibrations assume a definite form, the locus of the electric field lines, ellipses, or circles.

212

Our further assumption has been that homogeneous and stabilized glass can generally be considered as an isotropic amorphous material. If, however, glass is internally stressed, it becomes optically anisotropic and manifests double refraction.

Interference of radiation occurs if two waves vibrating with equal amplitudes add up geometrically (vector addition) to yield a single resulting wave. The interfering waves then mutually reinforce, attenuate, or neutralize each other. This chapter will discuss these phenomena.

4.5.2   Polarization of radiation

Owing to the fact that radiation is the result of a number of phenomena produced by different oscillators, the intensity planes of the electric and magnetic components do not adopt constant positions: they rotate arround the direction of propagation.

If, however, the vibrations maintain a particular arrangement, the radiation is then said to be polarized: plane-polarized, elliptically polarized, or circularly polarized [1]. Since the intensity planes of the electric and magnetic fields (see Chapter 3) are mutually perpendicular, we can take advantage of this fact in our discussion of optical effects by considering solely the orientation of the vector of the electric wave.

If the vibrations are arrayed in a single plane, we have the case of the *plane-polarized radiation*, which is shown in Figure 159.

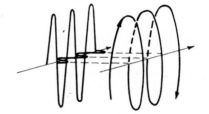

Fig. 159 Plane polarized radiation.

Fig. 160 Eliptically pollarized radiation. ▶

Elliptically polarized radiation is the result of a combination of two plane-polarized radiant flows with the directions of their vibrations mutually orthogonal. The tip of the electric vector then draws a line in the form of a helix with an elliptic base, the shape of which depends upon the phase difference between the plane-polarized rays. Figure 160 presents a schematic diagram of elliptically polarized radiation.

If the amplitudes of the two plane-polarized rays are equal and the path difference $\varDelta$ equals $\frac{\lambda}{4} (2k - 1)$ for $k = 1, 2, 3, \ldots$, we get *circularly polarized radiation*. The tip of the radiation vector then draws a line in the form of a helix with a circular base. Figure 161 presents a schematic diagram of circularly polarized radiation.

Polarization by reflection

Radiation incident upon the interface between two media of different indices of refraction is partially reflected and partially refracted. The amount of radiation reflected depends upon the angle of incidence, Θ_1; the larger the angle of incidence,

Fig. 161 Circularly polarized radiation.

the larger the reflected part of the radiation. If the plane of vibration of the incident radiation is normal to the plane of incidence, we can write for the amplitude of the reflected radiation, A_s'' [2],

$$A_s'' = -A_s \frac{\sin(\Theta_1 - \Theta_2)}{\sin(\Theta_1 + \Theta_2)}, \tag{237}$$

where A_s'' is the amplitude of the reflected radiation,
$\quad A_s$ is the amplitude of the incident radiation,
$\quad \Theta_1$ is the angle of incidence,
and $\quad \Theta_2$ is the angle of refraction.

If the intensity of the radiant flux is proportional to the second power of the amplitude, then

$$\perp \varrho_s = \frac{\Phi_s''}{\Phi_s} = \left(\frac{A_s''}{A_s}\right)^2 = \frac{\sin^2(\Theta_1 - \Theta_2)}{\sin^2(\Theta_1 + \Theta_2)}, \tag{238}$$

where Φ_s'' is the intensity of the radiant flux reflected,
and $\quad \Phi_s$ the intensity of the incident radiant flux.

If the plane of vibration of the incident radiation is parallel to the plane of incidence,

$$\| \varrho_p = \frac{\Phi_p''}{\Phi_p} = \left(\frac{A_p''}{A_p}\right)^2 = \frac{\tan^2(\Theta_1 - \Theta_2)}{\tan^2(\Theta_1 + \Theta_2)}. \tag{239}$$

If the plane of vibration of the incident radiation and the plane of incidence enclose a certain angle, the radiation can be split into two components; one component normal to the plane of incidence, the other parallel to it. Complex radiation is a mixture of waves polarized in all possible planes. Each of these can be split into two components, viz. perpendicular, $\perp$, and parallel, $\|$, to the plane of incidence. The values of the two components are equal, so that for complex radiation we can write

$$\varrho = \frac{1}{2}(\perp \varrho_s + \| \varrho_p). \tag{240}$$

214

By selecting the proper angle of incidence, Θ_1, we can ensure that

$$\Theta_1 + \Theta_2 = \frac{\pi}{2},$$

so that $\| \varrho_p = 0$, and

$$\varrho = \frac{1}{2} \perp \varrho_s. \tag{241}$$

In this case only the component perpendicular to the plane of incidence is reflected, and the radiation is plane-polarized. We can then write, after Brewster, that

$$n \sin \Theta_1 = n' \sin\left(\frac{\pi}{2} - \Theta_1\right), \tag{242}$$

thus

$$\tan \Theta_1 = \frac{n'}{n} = \tan \Theta_p, \tag{243}$$

where n is the refractive index of the glass
and n' is the refractive index of the other medium.

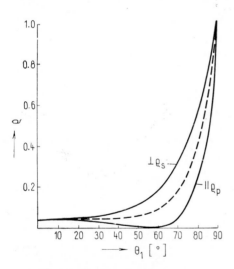

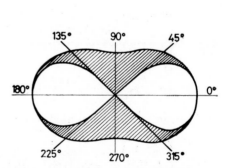

Fig. 162 Dependence of reflectivity upon the angle of incidence, Θ_1, for perpendicular ($\perp \varrho_s$) and parallel ($\| \varrho_p$) components.

Fig. 163 Radiation polarized by scattering from Rayleigh type scattering centres. Hatched area visualizes polarized radiation.

The angle $\Theta_1 = \Theta_p$ in equation (243) is the *polarizing angle* and the relationship is termed *Brewster's law*. Figure 162 shows the dependence of reflectance upon the angle of incidence, Θ_1, for the perpendicular component, $\perp \varrho_s$, and parallel component, $\| \varrho_p$.

Polarization by refraction

In this case the vibration plane is parallel to the plane of incidence, and for the amplitudes $\perp A'_s$ and $\| A'_p$ of the refracted radiation we can write [2]

$$\perp A'_s = A_s \frac{2 \cos \Theta_1 \sin \Theta_2}{\sin (\Theta_1 + \Theta_2)},$$ (244)

$$\| A'_p = A_p \frac{2 \cos \Theta_1 \sin \Theta_2}{\sin (\Theta_1 + \Theta_2) \cos (\Theta_1 - \Theta_2)}.$$ (245)

The polarizing effect is enhanced by an increased number of refractions: the ratio $\dfrac{\| A'_p}{\perp A'_s}$ and the magnitude of the polarized component $\| A'_p$ will increase accordingly.

For a plate of glass with two surfaces we can write for refraction at the first surface

$$\perp A'_{s1} = \frac{2 \cos \Theta_1 \sin \Theta'_1}{\sin (\Theta_1 + \Theta'_1)},$$ (246)

$$\| A'_{p1} = \frac{2 \cos \Theta_1 \sin \Theta'_1}{\sin (\Theta_1 + \Theta'_1) \cos (\Theta_1 - \Theta'_1)};$$ (247)

and at the second surface

$$\perp A'_{s2} = \perp \frac{2 \cos \Theta_2 \sin \Theta'_2}{\sin (\Theta_2 + \Theta'_2)},$$ (248)

$$\| A'_{p2} = \| \frac{2 \cos \Theta_2 \sin \Theta'_2}{\sin (\Theta_2 + \Theta'_2) \cos (\Theta_2 - \Theta'_2)}.$$ (249)

Since

$$\Theta_2 = \Theta'_1$$
$$\Theta'_2 = \Theta_1$$ (250)

the following relationship holds for m plates:

$$\left(\frac{\| A'_{p2}}{\perp A'_{s2}} \right)_m = \frac{1}{\cos^{2m}(\Theta_1 - \Theta'_1)}.$$ (251)

Polarizing by scattering

Scattering centres polarize radiation perpendicularly to the plane given by the incident radiant flux and by the direction of observation. The intensity of scattered light depends upon the angle of the observation. The proportion of unpolarized radiation at a given angle of observation can be read from the internal indicatrix in Figure 163; it can be seen that the radiation scattered in the 0° or 180° directions is unpolarized, and radiation scattered at 90° is fully polarized.

216

4.5.3 Double refraction in glasses

We invariably assume that glass is an isotropic material, so that no double refraction (birefringence) of radiation passing through it can occur.

If, however, it manifests *internal stress*, glass can be viewed as an anisotropic material that causes birefringence (see Figure 164).

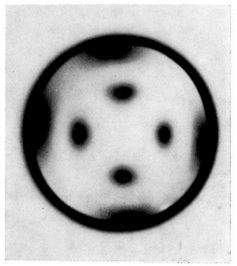

Fig. 164 Birefringence in internally stressed glass (*a*) compared with almost zero internally stressed glass (*b*).

Internal stress in glass can be produced by mechanical stress, thermal shock, inhomogeneities present in glass, phase separation, crystallization, metal precipitation, etc. [3, 17].

If glass is strained in tension, the index of refraction will decrease in a direction parallel to the tension and it will increase in a direction normal to the tension. Subjecting glass to compressive stress will increase its index of refraction in a direction parallel to that of the stress, whereas in a direction normal to it, the refractive index will decrease.

Under compressive stress, glass behaves, in an optical sense, as a negative uniaxial crystal. On the other hand, under tensile stress it becomes a positive uniaxial crystal. The optical axis then coincides with the direction of the stress.

If light passes through glass that has internal stress in a direction different from the optical axis, *double refraction* can be observed. The peak value will be reached if the direction of propagation is normal to the optical axis. The light then splits into two parts, *ordinary* and *extraordinary*, which pass through the glass at different velocities. These velocities are inversely proportional to the indices of refraction n_O and n_E (where the subscript O stands for ordinary, subscript E for extraordinary). The path difference after passage through the glass is equal to the difference in optical paths. For the optical path we can write

$$\vartheta' = n\vartheta, \tag{252}$$

where ϑ' is the optical path in the medium of refractive index n,

n is the index of refraction of the glass,

and ϑ is the actual length of the path corresponding to the thickeness of the glass.

The difference (Δ) in optical path between ordinary and extraordinary rays is

$$\Delta = (n_O - n_E)\vartheta. \tag{253}$$

Birefringence is then proportional to the stress viz.

$$D = n_O - n_E) = B\sigma, \tag{254}$$

where D is the birefringence of the glass,

σ is the stress in the glass,

and B is the photoelastic constant.

The magnitude of the photoelastic constant depends upon the chemical composition of the glass, and its dimension is $[m . s^2 . kg^{-1}]$.

Substituting into equation (253) we get

$$\Delta = B\sigma\vartheta. \tag{255}$$

Stress in glasses is usually measured with the help of the polarization equipment, which can be classified into two groups: *polariscopes* and *polarimeters*.

A polariscope is an instrument used for examining glass in polarized light, and the path difference is ascertained by the type of interference colour, or determined with some approximation with the help of comparators (or compensators).

A polarimeter has an inbuilt compensator with which the path difference can be measured. The principal parts of the apparatus are a light source, a polarizer, an analyzer, possibly a screen, a sensitive interference plate, a quarter-wave plate, a colour filter, and a compensator. Experimental techniques to measure the birefringence are adequately described in various textbooks on stress optical effect; an ASTM standard (C148 − 59T) was also established [17].

For details the reader is referred to ref. [3, 17], particularly to chapters dealing with annealing and the control of stress in glasses.

4.5.4 Polarization by absorption

A number of birefringent crystals, e.g., tourmaline, have the property of absorbing one of the rays of the double refracted beam of light. In tourmaline, the ordinary ray is absorbed and the extraordinary ray, partially attenuated, passes through. The absorption is selective, so that the radiation transmitted is chromatic. Tourmaline thus exhibits so-called *dichroism*.

218

Crystals manifesting these properties are employed in the construction of polarizing equipment. In the production of polarization filters use has mostly been made in the past years of foils of plane-polarized macromolecules of polyvinyl alcohol coloured with iodine or reduced metal.

4.5.5 Rotary polarization

A number of optically active materials, e.g., crystalline quartz, manifest the property of rotating the plane of the direction of the vibrations of plane-polarized radiation in the direction of their optical axis. Some materials rotate the plane of polarization to the right (dextro-rotatory substances), others to the left (laevo-rotatory substances).

The rotatory capacity of quartz is based upon the distribution of the silicon and oxygen atoms along the optical axis. According to Fresnel, rotary polarization can be interpreted as follows.

Let us consider the state of polarized radiation incident on the optically active plate with the direction of the vibrations parallel to axis x, viz.

$$x = A \sin \omega t. \tag{256}$$

This vibration can be substituted by two vibrations, x_1 and x_2, of the same direction in the polarizing amplitude, viz.

$$x_1 = \frac{1}{2} A \sin \omega t; \qquad x_2 = \frac{1}{2} A \sin \omega t. \tag{257}$$

Two more vibrations must be added, viz. y_1 and y_2, in the direction of axis y exhibiting the same frequency and amplitude, but opposite phase:

$$y_1 = \frac{1}{2} A \sin\left(\omega t - \frac{\pi}{2}\right) = -\frac{1}{2} A \cos \omega t,$$
$$y_2 = \frac{1}{2} A \sin\left(\omega t + \frac{\pi}{2}\right) = \frac{1}{2} A \cos \omega t. \tag{258}$$

Thus we get two circularly polarized vibrations of equal amplitude. Fresnel assumed that each of these vibrations propagates through the crystal at a different velocity, so that after covering path l, they will manifest a certain path difference, Δ, viz.

$$\Delta = (n_2 - n_1)l, \tag{259}$$

where n_1, n_2 are the indices of refraction corresponding to the individual vibrations.

For the resulting vibrations after path l has been covered, we can write

$$x' = x_1' + x_2' = \frac{1}{2} A[\sin \omega t + \sin(\omega t - \varphi)] =$$
$$= A \cos \frac{\varphi}{2} \sin\left(\omega t - \frac{\varphi}{2}\right), \tag{260}$$

$$y' = y_1' + y_2' = \frac{1}{2} A[(\cos \omega t - \varphi) - \cos \omega t] =$$

$$= A \sin\left(\omega t - \frac{\varphi}{2}\right) \sin \frac{\varphi}{2},$$

where

$$\varphi = \frac{2\pi}{\lambda} \varDelta. \tag{261}$$

The vibration is thus rotated through an angle α, where

$$\tan \alpha = \frac{y'}{x'} = \tan \frac{\varphi}{2}, \tag{262}$$

so that

$$\alpha = \frac{\varphi}{2} = \frac{\pi}{\lambda} l(n_2 - n_1). \tag{263}$$

4.5.6   Interference of radiation

If two sources of light, z_1 and z_2, emit monochromatic waves, these waves may interfere and the intensity of the light will therefore change [4].

If the intensity of the radiation is proportional to squared amplitude.

$$\varPhi = KA^2, \tag{264}$$

we can write for the resulting squared amplitude

$$A^2 = A_1^2 + A_2^2 + 2A_1A_2 \cos (\varphi_2 - \varphi_1), \tag{265}$$

where φ_1, φ_2 denote the phase shifts.

The value of the intensity of the resulting wave will thus be dependent upon the value of $\cos (\varphi_2 - \varphi_1)$.

If $\cos (\varphi_2 - \varphi_1) = 1$, the intensity of the resulting wave will be maximal, viz.

$$\varPhi_{max} = KA^2{}_{max} = K(A_1 + A_2)^2. \tag{266}$$

For the intensity to reach this peak, with $\cos (\varphi_2 - \varphi_1) = 1$, it is thus essential that the phase shift be

$$\delta = \varphi_2 - \varphi_1 = 2m\pi, \tag{267}$$

where $m = 0, 1, 2, \ldots$.

The path difference $\varDelta$ will then, under the condition of maximum intensity, be

$$\varDelta = \varDelta_2 - \varDelta_1 = m\lambda, \tag{268}$$

where $\varDelta_1$, $\varDelta_2$ denote the paths of the two rays.

The intensity of the resulting wave will reach a minimum if $\cos (\varphi_2 - \varphi_1) = -1$. Then,

$$\varPhi_{min} = KA_{min}^2 = K(A_1 - A_2)^2. \tag{269}$$

The phase difference, δ, will then be

$$\delta = \varphi_2 - \varphi_1 = 2m\pi, \tag{270}$$

where $m = 1/2, 3/2, 5/2, \ldots,$
and the path difference

$$\Delta = m\lambda. \tag{271}$$

The interference order of the respective maximum or minimum is denoted by the letter m. The intensity of the light is thus amplified at places where the waves meet, with path difference being an integer multiple of the wavelength. It decreases, or drops to zero, at places where the path difference is equal to an odd multiple of a half wavelength.

For interference effects to be observable, the source of light must be coherent, viz. it must be capable of sending out waves the phase difference of which is constant. If the human eye can sense changes in the intensity lasting at least 0.1 sec, it is quite understandable that sources with a constant path difference as low as 10^{-5} (sodium vapour lamps) should be considered unsuitable for the observation of interference. Hence, use is invariably made of a single source, the light from which is split into two waves by division of wave front or amplitude.

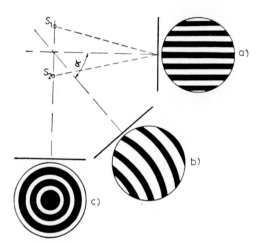

Fig. 165 Shape of interference fringes dependent upon the position of the shade.

In the past few years laser light sources have increasingly been employed. The radiation that these sources emit is phase-constant for $0.01-0.1$ sec (occasionally even longer).

Adjusting the position of the screen on which the interference figures can be observed, we obtain the characteristic shapes of the interference bands (Figure 165).

The effect of interference of light is utilized in the examination of the quality of glasses. This method makes it possible to measure the refractive index, wavelengths, fine structure of spectrum lines, thicknesses of thin layers, etc.

The interference method of measuring the thickness of thin films is based upon the *multi-ray interference of radiation*. After applying the film we see that an abrupt transition is produced between the uncoated part of the surface of the glass and part coated with the layer. If then the layer is viewed perpendicularly in monochromatic light, we observe interference effects, which are characterized by light and dark interference fringes of equal thickness.

If dy is the distance between two parallel interference bands, which corresponds to a change in the thickness of $\lambda/2n_1$, then the displacement of the y bands caused by depression l is also proportional to the value of l, viz. to the thickness of the film layer sought. For l we can then write

$$l = \frac{y}{dy} \frac{\lambda}{2n_1}.$$ (272)

Measuring the displacement of the bands at the point of depression, y, and the distance between two adjoining bands, dy, we determine the thickness for the given wavelength and refractive index of the layer.

Fig. 166 Interference fringes in a thin reflecting layer of glass after etching.

For white light the thickness of the layer, l, may be computed from the relationship

$$l = \frac{1}{2n_1} (m'\lambda'_m - m\lambda_m),$$ (273)

where m, m' are the interference orders obtained from the relationships

$$m = \frac{\lambda_2}{\lambda_1 - \lambda_2}; \qquad m' = \frac{\lambda'_{m+1}}{\lambda'_m - \lambda'_{m+1}},$$

where λ_m, λ_{m+1} are the wavelengths of two neighbouring interference minima, and λ'_m is the wavelength of the corresponding interference minima for the radiation reflected at the point of depression.

In Figure 166 we present the interference figure of a thin reflecting layer at the point of depression.

The interference method can also be used for checking the flatness of surfaces and their microstructures, as shown in Figure 167, which shows the structure of a chemically polished surface of glass.

The principle of the interference effect has also found application in the production of reflecting layers and antireflection coatings, interference filters, etc. This subject is dealt with in more detail in Chapter 7.

Fig. 167 Interference fringes observable in a well chemically polished (a) and a badly polished (b) surface of glass.

The instruments used for measuring the interference of radiation are called *interferometers*. The most widely known are the ones designed by Jamin, by Michelson, by Fabry-Perot and by others. The interferometer can be used in combination with a microscope, where the interference pattern will show in the enlarged image of the specimen examined. This apparatus may be used both in reflected and in transmitted light. For more details the reader is referred to refs. [4, 8].

References

[1] V. Prosser, Optické vlastnosti pevných látek (Optical Properties of Solid Materials). SPN, Prague, 1971.

[2] B. Havelka, E. Keprt, and M. Hansa, Spektrální analýza (Spectral Analysis). NČSAV, Prague, 1957.

[3] F. Schill, V. Novotný, and Z. Hrdina, Chlazení skla a kontrola pnutí (Annealing of Glass and Stress Control). SNTL, Prague, 1968.

[4] J. Sládková, Interference světla (Interference of Light). SNTL, Prague, 1967.

[5] V. Novotný and F. Schill, Měření pnutí ve skle Sénarmontovým kompenzátorem Meopta (Measuring stress in glass with the help of MEOPTA Sénármont compensator). *Sklář a keramik*, *11* (1961), 303—307.

[6] M. Tashiro, The effect of polarisation of the constituent ions on the photoelastic birefringence of the glass. *Soc. Glass Technol.*, *40* (1956), pp, 353—362.

[7] M. Milbauer and M. Perla, Fotoelasticimetrické přístroje a měřicí metody (Photoelastic Apparatus and Methods of Measurement). ČSAV, Prague, 1959.

[8] J. Brož et al., Základy fyzikálních měření (Foundations of Physical Measurement). SPN, Prague, 1974.

[9] A. Vašíček, Optics in Thin Films. North Holland Publ. Co., Amsterdam, 1960.

[10] W. Krug, J. Rienitz, and G. Schulz, Beiträge zur Interferenzmikroskopie. Akademie-Verlag, Berlin, 1961.

[11] J. Čajko, Registračný vysokorozlišovací spektrometr s Fabryovým-Perotovým interferometrem (High resolution recording spectrometer with Fabry-Perot interferometer). *Čs. čas. fyz.*, *14* (1964) 105.

[12] M. Miller, Holografie (Holography). SNTL, Prague, 1974.

[13] F. T. S. Yu, Introduction to Diffraction, Information Processing, and Holography. The MIT Press, Cambridge, Mass., 1973.

[14] M. Garbuny, Optical Physics. Academic Press, New York—London, 1965.

[15] L. Eckertová, Fyzika tenkých vrstev (Physics of Thin Layers). SNTL, Prague, 1973.

[16] A. Štrba, Všeobecná fyzika 3 — optika (General Physics 3 — Optics), Alfa — SNTL, Bratislava—Prague, 1979.

[17] M. Tomozawa, R. H. Doremus, Treatise on Materials Science and Technology, Vol. 12. Glass I.: Interaction with Electromagnetic radiation. Anomalous Birefringence in Oxide Glasses (T. Takamori, M. Tomozawa), Academic Press, New York, San Francisco, London 1977, pp. 123—155.

5 CLASSIFICATION OF COMMERCIALLY PRODUCED GLASSES IN TERMS OF OPTICAL PROPERTIES

5.1 INTRODUCTION

The majority of technical, utility, luxury and imitation jewellery glasses, colourless or coloured, are standardized as far as their physico-chemical properties are concerned.

Very often, the optical properties are of utmost importance from the point of view of both the appearance and the application of the respective glass products. The range of commercially produced glasses is rather large, our classification will be thus selective so as to cover only the types whose optical properties are of decisive importance.

5.2 COLOURLESS GLASSES

5.2.1 Technical glasses

Optical and special optical glasses

Among the colourless technical glasses the most important are the optical and special optical glasses.

These glasses are widely used in the production of optical instruments and optical equipment, and the quality criteria applied to them are invariably very high. As already mentioned in the preceding chapters, the optical properties of glasses depend on the chemical composition. In order to meet the rising demands of the optical industry, a large palette of optical glasses has been developed of widely differing optical properties, as shown in Figure 168.

In the $n_d - v_d$ diagram, optical glasses are classified into several groups denoted by letters, the individual glasses by numbers. In Table 37 we present examples of the denotation of optical glasses in the Czechoslovak, German, Soviet, and French nomenclatures.

Some examples of the composition and optical properties of glasses are given in Table 38.

Optical glasses are (after [1]) classified by deviation of the refractive index n_d of the semifinished products from the rated value into four categories; and by homogeneity in the refractive index into six classes. From the point of view of

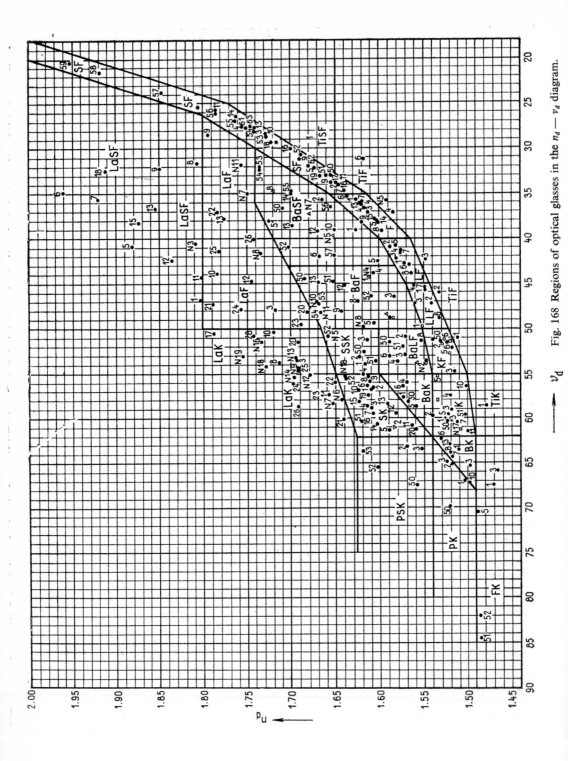

Fig. 168 Regions of optical glasses in the $n_d - \nu_d$ diagram.

Table 37

Group	CSSR	Schott GDR	USSR	Parra Mantois
Barium flint	BaF 557/486	BaF 1		FB B5648
	BaF 570/495	BaF 2	BF 6	B7749
	BaF 583/465	BaF 3		
	BaF 607/492	BaF 5		CO749
	BaF 589/486	BaF 6		
	BaF 608/462	BaF 7	BF 17	FB C1046
Dense barium flint	BaTF 626/391	BaSF 1	BF 12	FBC C2639
	BaTF 607/402	BaSF 3		C0740
	BaTF 603/425	BaSF 5		C0342
Crown flint	KF 526/510	KF 2	KF 3	CHD B3051
	KF 515/547	KF 3	KF 2	B1554
Extra light flint	LLF 548/549	LLF 1		FEL B4946
	LLF 541/472	LLF 2		B4148
	LLF 532/489	LLF 6		B3149
Light flint	LF 573/427	LF 1		FL B7349
	LF 579/417	LF 4		B8042
	LF 575/413	LF 7	LF 5	B7741
Flint	F 626/356	F 1		FD C2636
	F 620/363	F 2		C2036
	F 613/370	F 3	F 1	C1337
	F 617/366	F 4	F 2	C1737
	F 603/380	F 5	F 6	C0438
	F 636/354	F 6		C3735
	F 625/356	F 7	F 4	C2636
	F 595/392	F 8		B9639
Dense flint	TF 717/295	SF 1	TF 3	FED D1729
	TF 648/339	SF 2	TF 1	C4834
	TF 755/275	SF 4	TF 5	D5427
	TF 673/322	SF 5	TF 2	C7232
	TF 640/346	SF 7		
	TF 655/338	SF 9		C5333
	TF 648/338	SF 12		C4834
	TF 667/331	SF 19		C6733
Short flint	KrF 529/518	KzF 2		CHD B3051
Crown	K 510/619	K 1		
	K 516/568	K 2		C B1657
	K 518/590	K 3	KF 4	B1858
	K 523/596	K 5		B2359
	K 511/606	K 7		B1161
	K 513/598	K 8		B1359

Table 37 (cont.)

Group	CSSR	Schott GDR	USSR	Parra Mantois
	K 501/565	K 10		B0157
	K 522/591	K 13		B2359
Barium	BK 510/634	BK 1	K 3	BSC B1064
crown	BK 506/620	BK 2		
	BK 531/621	BK 6		C B3061
	BK 516/640	BK 7	K 8	BSC B1664
	BK 519/642	BK 12	K 17	B1864
Zinc	ZK 533/580	ZK 1		C B3358
crown	ZK 521/602	ZK 2		B1960
	ZK 534/554	ZK 5	K 15	B3556
Light	BaLK 526/601	BaLK 1	K 20	B2760
crown	BaLK 518/603	BaLK 3	L 18	B1860
barium				
Barium	BaK 572/575	BaK 1	TK 2	BCL B7457
crown	BaK 540/596	BaK 2	BK 6	B3959
	BaK 569/560	BaK 4	BK 10	B6956
Dense	BaTK 607/576	SK 2		BCD C0856
barium	BaTK 613/586	SK 4	TK 16	C1258
crown	BaTK 589/612	SK 5	TK 3	B8761
	BaTK 614/563	SK 6	TK 7	C1356
	BaTK 623/569	SK 10	TK 20	C2157
	BaTK 620/603	SK 16	TK 1	C2060
	BaTK 617/540	SSK 1	TK 9	C1754
	BaTK 622/531	SSK 2	BF 11	C2353
	BaTK 658/508	SSK 5	TK 21	C5851
Light	BaLF 571/530	BaLF 3		BCD B7052
barium	BaLF 580/539	BaLF 4	BF 7	B8253
flint	BaLF 547/536	BaLF 5		B4853
	BaLF 589/532	BaLF 6		B8753
	BaLF 589/511	BaLF 7	BF 19	

mean dispersion, $n_F - n_C$, optical glasses are grouped into four categories, and by their homogeneity in mean dispersion, $n_F - n_C$, into four classes.

The classification of optical glasses in terms of birefringence has five categories. The other criteria are: the degree of optical homogeneity, striation, seediness and, above all, absorption (classification into six categories).

The development of atomic energy has established the need for protection of workers processing radioactive isotopes. The inspection ports are glazed with optical glass of high lead monoxide concentration. These glasses are characterized by the so-called *lead equivalent*, which, for example, for the TF 755/275 glass equals 2.44; this indicates the protective capacity of the glass as compared with the protective capacity of a lead plate. This subject has been discussed in detail by Brewster [17].

Since glasses with a high content of lead monoxide generally have a low chemical durability, they must be protected by one, or several, glasses manifesting higher chemical durability, e.g., by the BK 516/640 glass. This glass must, however, be stabilized, viz. immunized against the effects of the radiation, by introducing into it, for example, 0.5 wt. % of ceric oxide. In order that losses of light due to reflection may be reduced, the surfaces of the inspection ports must be coated with an antireflection layer.

The groups of glasses in the $n_d - v_d$ diagram that have extreme optical properties are referred to as *special optical glasses*. They are capable of satisfying the exacting requirements of the designers of optical systems. For instance, lanthanum glasses with an addition of thorium oxide, tungsten trioxide, tantalum pentoxide, niobic pentoxide, of a high refractive index and relatively high Abbe value, enable a fairly good suppression of the chromatic, and a more effective correction of the spherical, aberration across the whole diameter of the lens. Apart from this, lanthanum glasses make it possible to reduce distortion of the image and the astigmatism of an optical system for an enlarged viewing field. These glasses form the basis of most of the fast wide-angle lenses, whose definition of image is superior to that of lenses of lower parameters. In Table 39 we give the composition and the properties of several types of special optical glasses.

In order to perfect the design of the microscope lens, use can be made of glasses with a low index of refraction, n_d, but high Abbe value, v_d, the so-called *fluoride* or *phosphofluoride glasses*. Owing to their greatly limited dispersion, these glasses permit the construction of apochromatic and plano-apochromatic lenses. Examples of the composition of the quoted types of glasses and their properties are given in Table 40.

Semi-optical glasses

This group of technical (colourless) glasses comprises *spectacle glasses*. Despite the fact that they are classified with this group, the parameters of these glasses can very often compare with the parameters of optical glasses. The range of shapes and dimensions of the blanks is very large and it amounts to several hundred types. From the point of view of optical properties, these glasses are defined particularly by the index of refraction ($n_d = 1.523$), a high degree of homogeneity, and numerous other properties as, for example, the retention of the curvature of the pressing etc.

The blanks differ from each other in their diameter and thickness, and in the curvature of their surfaces. The dioptric value of the pressings ranges between 0 to ± 20 dioptres, and the most frequently marketed are the ± 6 dioptre values, the lowest dioptrical difference equalling 0.5.

Table 38

Glass	Composition [wt. %]											
	SiO_2	B_2O_3	Al_2O_3	CaO	ZnO	BaO	PbO	K_2O	Na_2O	As_2O_3	Sb_2O_3	
K 523/596	62.5	2.0			3.0	11.0	1.0	15.0	5.0	0.5		$n_d = 1.5225$ $v_d = 59.6$ $n_F - n_C = 0.008\,76$
BK 516/640	68.9	10.1				2.8		8.4	8.8	0.5	0.5	$n_d = 1.5163$ $v_d = 64$ $n_F - n_C = 0.008\,06$
ZK 534/554	58.2	3.1	1.0		8.7	1.5	5.4	8.0	13.8	0.3		$n_d = 1.5338$ $v_d = 55.4$ $n_F - n_C = 0.009\,64$
BaLK 518/603	68.0	5.0			0.1	7.7		4.1	13.8	0.4		$n_d = 1.5183$ $v_d = 60.3$ $n_F - n_C = 0.008\,59$
BaK 569/560	50.4	5.9			11.5	19.7	2.0	5.0	4.0	1.0	0.5	$n_d = 1.5688$ $v_d = 56$ $n_F - n_C = 0.010\,15$
BaTK 620/603	30.8	17.9	1.4			48.7			0.3	0.5	0.4	$n_d = 1.620\,41$ $v_d = 60.3$ $n_F - n_C = 0.010\,29$
BaTK 658/508	30.1	10.1	0.7	3.5	5.1	46.7				0.5	0.3	$n_d = 1.658\,44$ $v_d = 50.8$ $n_F - n_C = 0.012\,95$

Table 38 (cont.)

Glass	Composition [wt. %]											
	SiO₂	B₂O₃	Al₂O₃	CaO	ZnO	BaO	PbO	K₂O	Na₂O	As₂O₃	Sb₂O₃	
BaLF 541/536	57.2				8.0	14.0	6.7	11.0	2.0	0.8	0.1	$n_d = 1.5474$ $v_d = 53.6$ $n_F - n_C = 0.010\ 21$
BaF 589/486	48.6	1.0			8.4	20.5	14.2	6.3	0.5	0.5		$n_d = 1.5890$ $v_d = 48.6$ $n_F - n_C = 0.012\ 11$
BaTF 607/402	46.3				3.7	8.0	33.0	8.0	0.5	0.5		$n_d = 1.6072$ $v_d = 40.2$ $n_F - n_C = 0.015\ 09$
LLF 541/472	63.2						23.5	8.0	5.0	0.3		$n_d = 1.5407$ $v_d = 47.2$ $n_F - n_C = 0.011\ 45$
LF 575/413	53.9						34.9	7.9	2.1	0.8		$n_d = 1.5750$ $v_d = 41.3$ $n_F - n_C = 0.013\ 92$
F 620/363	45.7						45.1	5.0	3.6	0.6		$n_d = 1.6200$ $v_d = 36.3$ $n_F - n_C = 0.017\ 06$
TF 717/295	20.3						79.0	0.4		0.3		$n_d = 1.717\ 36$ $v_d = 29.5$ $n_F - n_C = 0.024\ 31$
KrF 529/518	49.0	20.0	2.6					6.1	2.1	1.0	20.0	$n_d = 1.529\ 44$ $v_d = 51.8$ $n_F - n_C = 0.010\ 20$

Table 39

	Composition [wt. %]							n_d	v_d
	B$_2$O$_3$	La$_2$O$_3$	Ta$_2$O$_5$	ThO$_2$	ZrO$_2$	SiO$_2$	CdO		
1	29.0	42.5	13.5	10.0	5.0	—	—	1.799 91	40.3
2	19.0	30.0	11.0	10.0	—	5.0	25.0	1.821 15	42.3
3	19.0	22.5	11.0	15.0	—	5.0	27.5	1.820 92	42.1
4	28.5	38.0	16.0	15.0	2.5	—	—	1.798 75	45.4
5	25.0	37.5	2.5	—	5.0	5.0	25.0	1.783 38	45.8
6	33.5	41.5	7.5	12.5	5.0		—	1.770 67	48.8

Table 40

	Composition [wt. %]								n_d	v_d
	BeF$_2$	AlF$_3$	KF	MgF$_2$	CaF$_2$	SrF$_2$	BaF$_2$	LaF$_3$		
1	35.3	23.0	—	9.0	11.3	5.4	7.5	8.5	1.356 73	101.2
2	46.8	14.1	7.7	10.3	10.4	4.2	—	6.5	1.334 81	105.2

In recent years spectacle glasses have been developed manifesting a high index of refraction ($n_d \geq 1.700$) and low specific mass ($\varrho \leq 3000$ kg/m^3). These glasses, with a high content of titanium dioxide, are used in the production of highly dioptric glasses of low mass.

5.2.2 Utility and luxury glasses

The production of utility and luxury glass has a long tradition in Czechoslovakia, and the quality and exclusiveness of Czechoslovak glass is known throughout the world. The term *"crystal glass"* is applied to glasses used in the production of utility and decorative objects of various shapes, plain or decorated [2] by various techniques.

Utility crystal glasses are classified by their properties into five groups, as shown in Table 41.

The optical properties are thus the decisive criterion for this classification. And, accordingly, the highest parameters are manifested by full lead and lead crystal glass, the lowest parameters by soda potash glass. This classification is in accordance with foreign standards [3, 4, 5, 6].

Utility and luxury glass may be ground, mechanically or chemically polished, engraved, sand-blasted, frosted, acid-embossed, obscured, stained, painted, etc.

For further details see refs. [7, 8].

Table 41

Groups	Important physical and chemical properties				
	Content of metallic oxides [wt. %]	Density [kg/m³]	Refractive index [n_d]	Microhardness HM 50	Durability [CSN 700 531]
1. Crystal glass soda potash	K_2O, CaO $\geq$ 10 $Fe_2O_3 \leq$ 0.025	—	—	—	IV/98 °C
2. Crystal glass	K_2O, BaO, PbO $\geq$ 10 $Fe_2O_3 \leq$ 0.025	$\geq$ 2400	—	550 $\pm$ 20	IV/98 °C
3. Special crystal glass	K_2O, ZnO, BaO, PbO $\geq$ 10 $Fe_2O_3 \leq$ 0.025	$\geq$ 2450	$\geq$ 1.520	—	IV/98 °C
4. Lead crystal	PbO $\geq$ 24 $Fe_2O_3 \leq$ 0.020	$\geq$ 2900	$\geq$ 1.545	— —	IV/98 °C
5. Full lead crystal	PbO $\geq$ 30 $Fe_2O_3 \leq$ 0.020	$\geq$ 3000	1.545	— —	IV/98 °C

5.3 COLOURED GLASSES

5.3.1 Technical glasses

Optical glass filters

In the group of coloured technical glasses the filter glasses are of great importance, particularly as far as the quality (homogeneity) and retention of precisely defined spectral characteristics are concerned. In technical practice these glasses are used for filtering complex radiation, in colour photography, and in special equipment.

These glasses, like optical glasses, are denoted by letters and numbers, which refer to precise spectral characteristics corresponding to catalogue values. In Table 42 the reader will find an example of how the various types of optical filtering glasses are classified according to German, Czechoslovak, and Soviet terminology.

Figure 169 presents the spectral characteristics of selected types of filter glasses, and Table 43 gives their compositions.

Apart from the optical filter glasses quoted, this group of special coloured glasses comprises fog glasses, coloured, for example, by the CdS + S combination (with Se added). These yellow glasses transmit radiation of wavelengths exceeding 500 nm. Since this (long-wave) radiation is more capable of propagating through fog and dust than short-wave radiation, this property is utilized especially in the production of fog lights for automobiles.

Table 42

Group of filters	Shade of colour	Code		
		GDR	CSSR	USSR
selective ultraviolet	black	UG	H$_s$	UFS
violet	violet	UG	G	PS
blue	blue	BG	F	SS
green	green	VG	E	ZS (ZZS)
yellow	yellow	GG	D	ZS
orange	orange	OG	C	OS
red	red	RG	B	KS
selective infra-red	black	RG (UG)	A$_s$	IKS
heat absorbant	bluish green	BG	T	SZS
neutral	grey	NG	N	NS

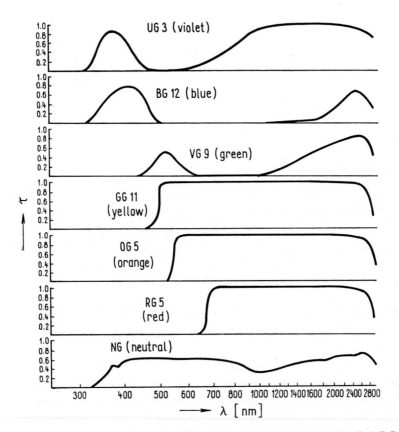

Fig. 169 Spectral characteristics of the UG 3, BG 12, VG 9, GG 11, OG 5, RG 5, and NG 5 filter glasses.

Table 43

Component	weight %						
	UG 3	BG 12	VG 9	GG 11	OG 5	RG 5	NG 5
SiO_2	50.0	70.0	58.5	62.0	66.0	60.0	60.0
B_2O_3	—	—	2.5	4.0	—	2.0	19.0
Al_2O_2	—	—	—	0.3	0.65	0.5	1.5
Na_2O	1.0	9.8	9.0	13.5	13.15	13.5	—
K_2O	13.2	10.0	9.0	10.0	8.0	6.0	12.0
CaO	—	—	—	10.0	—	—	—
ZnO	—	10.0	5.0	—	12.2	18.0	4.0
PbO	35.6	—	15.0	—	—	—	—
As_2O_3	0.2	0.2	1.0	0.2	—	—	—
KHF_2	—	—	—	—	—	1.0	5.0
cream of tartar	—	—	—	0.2	—	—	—
$KMnO_4$	3.0	—	—	—	—	—	—
CoO	—	1.43	—	—	—	—	2.02
CuO	—	0.427	0.85	—	—	—	—
$K_2Cr_2O_7$	—	—	1.0	—	—	—	—
CdS	—	—	—	0.2	1.2	1.0	—
Na_2SeO_3	—	—	—	—	—	1.0	—
S	—	—	—	0.3	—	—	—
Se	—	—	—	—	0.48	0.4	—
Fe_3O_4	—	—	—	—	—	—	2.25

Glasses of a blue tint, coloured by copper monoxide and cobalt monoxide are used in the production of electric bulbs simulating white daylight. The glass partially absorbs the red components from the glowing filament, so that the spectral energy distribution of the transmitted wavelengths corresponds to white daylight.

Of course, there are various other types of optical filter glasses; these are discussed in ref. [18].

Protective glasses

This group comprises sun glasses, welder's glasses, window glasses for reducing the glare of the sun, signal glasses and container glasses.

Sun glasses

In order to prevent our eyes from being dazzled and to ensure normal functioning even at increased luminance of the sun's radiation, we currently make use of protective sun glasses. Their spectral characteristics group them with protective filters, which are capable of limiting, or totally filtering off, a certain region of radiant energy.

The eye is exposed to three regions of radiation:

(a) to *visible radiation*, defined by the 380 nm – 780 nm waveband. The radiation of this region will not harm the eye, because the eye protects itself from being dazzled by contracting its iris;

(b) to *ultra-violet radiation*, defined by the wavelengths shorter than 380 nm. The harmful effect on the eye of this radiation is revealed only after several hours of exposure causing conjunctivitis and corneitis; it will not, however, result in long-lasting incapacity of the eye;

(c) to *infra-red radiation*, defined by the wavelengths exceeding 780 nm. The radiation of this region of the spectrum will cause coagulation of the lens protein, which will result in permanent injury to the eye — an eye disease termed "cataract".

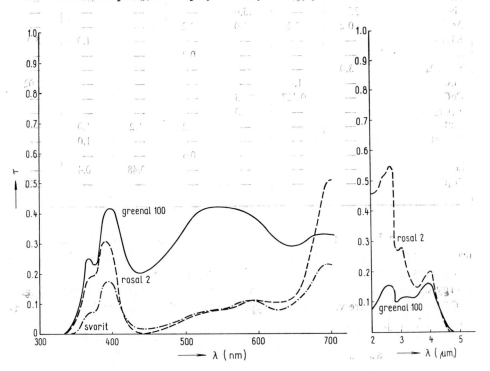

Fig. 170 Spectral characteristics of Czechoslovak Greenal 100, Rosal 2, and Svorit sun glasses.

Sun glasses can thus be defined as protective filters reducing the intensity of the harmful ultra-violet and infra-red regions of the radiation of the sun to a minimum and making the curve of the spectral characteristic in the visible region of the spectrum approximate the shape of the curve of the spectral sensitivity of the human eye. These requirements are best satisfied by the Czechoslovak Greenal sun glasses and, in part also, by the Svorit and Rosal sun glasses. In Figure 170 the reader will find plots of the spectral characteristics of these glasses.

For certain purposes (glass blowing, metallurgical operations, etc.) glasses are used containing, apart from other colourants, neodymium (Nd^{3+}). These glasses behave as selective filters of yellow colour, which is particularly advantageous for glass blowers. Figure 171 gives the curve of spectral transmission in protective glass of this type, and in Figure 172 the curves of spectral transmission in several other types of sun glasses are also given.

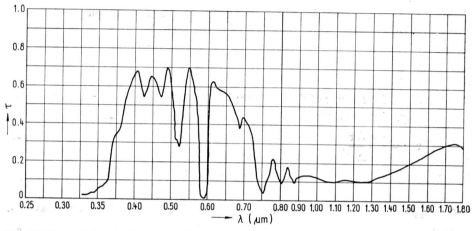

Fig. 171 Spectral characteristics of Nd^{3+} coloured protective glass.

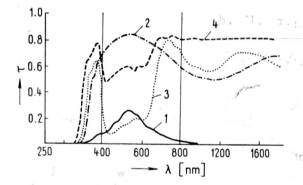

Fig. 172 Spectral characteristics of Sovirel sun glasses of:
1 — Dichrofil J9; *2* — Dichrofil 0.9;
3 — Brun F 18; *4* — Dichroroyal 08.

Sun glasses are not subject to specifications as far as their spectral characteristics are concerned, and production of these glasses is unfortunately governed by the whims of fashion.

Welder's glasses

The Czechoslovak standards CSN 70 7101 and 70 7103 define welding glasses by the degree of protection that these glasses are capable of rendering. The degree of protection is a numerical characteristic giving the level of protection from excessive and harmful ultra-violet and infra-red radiations and against too much glare in the visible region of the spectrum.

The degree of protection selected varies according to the different types of welding applied: flame welding, carbon electrode welding, metal arc welding or tungsten arc welding, etc. The degree of protection selected also depends on whether the glasses are only used as eye filters in metallurgical operations (lighter shades of colour) etc.

The calculation of the degree of protection is based on the measurement of spectral transmittance of the glass; glasses of lighter hues can be measured in their actual state, whereas glasses of darker hues must be abraded into thin plates and the results of the measurement converted into those corresponding to the original thickness.

In the *ultra-violet region* of the spectrum the transmission factor, $\tau_{(\lambda)}$, is measured continuously. For the λ wavelength it is calculated from the relationship

$$\tau_{(\lambda)} = \frac{(\Phi_{e\lambda})\,\tau}{\Phi_{e\lambda}}, \tag{274}$$

where $(\Phi_{e\lambda})\,\tau$ is the radiant flux of wavelength λ emerging from the specimen measured,

and $\Phi_{e\lambda}$ is the radiant flux of wavelength λ incident on the specimen measured.

In the *visible region* of the spectrum transmittance, τ_v, is measured continuously or discontinuously at 10 nm intervals. It is calculated from the relationship

$$\tau_v = \frac{\Phi_{\tau}}{\Phi} = \frac{\int\limits_{380\,\text{nm}}^{780\,\text{nm}} (\Phi_{e\lambda})_{\text{rel}} \cdot \tau_{(\lambda)} \cdot V_{(\lambda)} \cdot d\lambda}{\int\limits_{380\,\text{nm}}^{780\,\text{nm}} (\Phi_{e\lambda})_{\text{rel}} \cdot V_{(\lambda)} \cdot d\lambda}, \tag{275}$$

where Φ_{τ} denotes the radiant flux emerging from the specimen,

Φ the radiant flux incident on the specimen,

$(\Phi_{e\lambda})_{\text{rel}}$ the relative spectral energy distribution of the light source used (for the normal A white light at the 2 854 K colour temperature, if not stated otherwise),

$\tau_{(\lambda)}$ spectral transmittance of the specimen for wavelength λ,

and $V_{(\lambda)}$ relative luminous efficiency of monochromatic radiation (Czechoslovak standard CSN 01 1710).

The products of the spectral tristimulus values for the standard observer and the relative spectral energy distribution, $(\Phi_{e\lambda})_{\text{rel}}$, for the light source used are quoted in the Czechoslovak standard CSN 01 1718.

The current checks of transmittance in the visible region of the spectrum make use of the method of selected coordinates employing thirty selected coordinates.

In the *infra-region* of the spectrum the transmission factors τ_{NIR} and τ_{MIR} are calculated from the relationships

$$\tau_{\text{NIR}} = \frac{1}{520} \int_{780\,\text{nm}}^{1300\,\text{nm}} \tau_{(\lambda)}\, d\lambda, \tag{276}$$

$$\tau_{\text{MIR}} = \frac{1}{700} \int_{1300\,\text{nm}}^{2000\,\text{nm}} \tau_{(\lambda)}\, d\lambda. \tag{277}$$

If a plate whose thickness has been reduced by abrasion is subjected to measurement, the transmission factor must be converted so as to correspond to the original thickness, viz.

$$\tau_{l_1(\lambda)} = R \left(\frac{\tau_{l_2(\lambda)}}{R} \right)^{l_1/l_2}, \tag{278}$$

where $\tau_{l_1(\lambda)}$ is   the transmittance for the wavelength λ of the glass of the original thickness, l_1,

$R = \dfrac{2n}{n^2 + 1}$ the reflection factor for the two surfaces of the plane-parallel plate of the glass; the refractive index, n, is taken for the sodium, or helium, yellow spectrum line,

and $\tau_{l_2(\lambda)}$ the recorded transmission factor for the wavelength λ of the glass abrased to thickness l_2.

Spectral transmittance is measured at three points on the specimen at least. One point is central; the others are in the region of the 40 mm radius from the centre.

The differences in the values of spectral transmittance measured at various places on the same specimen must not exceed the error of measurement, M, allowed, as stated in the Czechoslovak standard CSN 70 7102. The calculation applies the relationship.

$$M = \frac{|\Delta\tau|}{\tau} \cdot 100 \quad [\%], \tag{279}$$

where $\Delta\tau$ is the difference between the values of transmittance recorded at two different places on the specimen,

and τ is the arithmetic mean of the values recorded.

According to the Czechoslovak standard CSN 70 7103, the computed values of spectral transmittance shall not exceed the values quoted in Table 44 for the given degree of protection.

In the 210−313 nm waveband the transmission factor shall not exceed the value permitted for the 313 nm wavelength, and in the 313−365 nm waveband that for the 365 nm wavelength. For visible radiation the value of the transmission factor for the 405 nm wavelength shall not exceed that for the 555 nm wavelength, and the transmission factor for the 666 nm wavelength shall still be below that for the 671 nm wavelength.

Table 44

Degree of protection N	Maximum value of transmittance in the ultra-violet region of the spectrum [%]		Transmittance in the visible region of the spectrum [%]		Maximum value of transmittance in the infra-red region of the spectrum [%]	
	$\lambda = 313$ nm	$\lambda = 365$ nm	from	to	τ_{NIR}	τ_{MIR}
3	0.000 3	2.8	17.8	8.5	12.0	8.5
4	0.000 3	0.95	8.5	3.2	6.4	5.4
5	0.000 3	0.30	3.2	1.2	3.2	3.2
6	0.000 3	0.10	1.2	0.45	1.7	1.9
7	0.000 3	0.037	0.45	0.17	0.81	1.2
8	0.000 3	0.031 3	0.17	0.060	0.43	0.68
9	0.000 3	0.004 5	0.060	0.023	0.20	0.39
10	0.000 3	0.001 8	0.023	0.008 5	0.10	0.25
11	0.000 3	0.000 6	0.008 5	0.003 2	0.050	0.16
12	0.000 2	0.000 2	0.003 2	0.001 2	0.027	0.098
13	0.000 076	0.000 076	0.001 2	0.000 45	0.014	0.060
14	0.000 027	0.000 027	0.000 45	0.000 17	0.007	0.040
15	0.000 009 4	0.000 009 4	0.000 17	0.000 06	0.003	0.020
16	0.000 003 4	0.000 003 4	0.000 06	0.000 023	0.003	0.020

Glasses giving protection against the radiation of the sun

Modern architecture has increasingly been making use of glass as a building material. It has therefore become indispensable to deal with the problems of the thermal losses and the regulation of thermal gains, in order that an acceptable measure of the interior temperature caused by the incident radiation of the sun may be obtained, or that losses in the interior temperature due to glazed walls may be reduced.

These problems are discussed by Kříž and Dubček [12, 13]. The authors survey the criteria applied to the properties of glasses giving protection against the effect of the sun's radiation; they also introduce the reader to the methodology of evaluation of these glasses. The protective effect is based either on the absorptivity or reflectivity of the glass, which are achieved either via the mass of the glass or by the application of coating layers.

Effective transmission, τ_{ef}, of the radiation of the sun is the first criterion. In the equilibrium state, τ_{ef} consists of regular transmission, τ_r, and of one part of the absorbed energy α_τ, as shown in Figure 173.

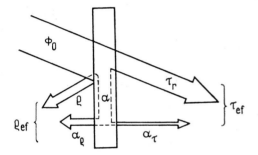

Fig. 173 Passage of solar radiation through glass.

The heat produced by the absorption, α, is transferred partly into the space enclosed, and partly into the space excluded by the glazing, α_τ and α_ϱ respectively. The ratio $\dfrac{\alpha_\tau}{\alpha_\varrho}$ depends upon the heat transfer coefficient on either side of the glass surface.

Considering glazing as a source of lighting, the transmission of visible radiation, τ_s, is the second criterion. The transmission of light should really be discussed in terms of effective transmission, since no glass exists that transmits luminous radiation only (i.e. no thermal radiation). This is why the so-called *selectivity factor* $F_s = \dfrac{\tau_s}{\tau_{ef}}$, is invariably quoted for protective glasses, which supplies us with the basic information about whether the glass is predominantly *thermally protective* ($F_s > 1$) or *luminously protective* ($F_s < 1$).

The transfer of energy into the interior of the buildings by means of diffused radiation is not of decisive importance and is dealt with under the same aspects as direct radiation.

241

The classification of protective glasses can, after [12, 13], be visualized in the way shown in Figure 174, where the values of τ_s are plotted against those of τ_{ef}.

In that part of the diagram to the left of the diagonal line $\tau_s = \tau_{ef}$ lie glasses showing mainly thermally protective properties; to the right of that line are glasses whose protection is mainly of a luminous nature. Region *1* are window glasses, region *2* determal (heat-absorbing) glasses (coloured with iron), region *3* grey glasses, bronze diathermals and Spectrofloat glass.

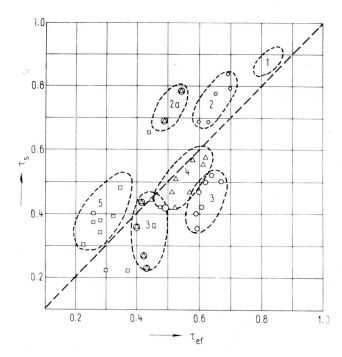

Fig. 174 Location of various types of protective glasses in the $\tau_s = f(\tau_{ef})$ diagram.

The dethermal glasses coated with dielectric layers lie in regions *2a* and *3a*. Colourless dielectric-coated glasses are grouped in region *4*, and vacuum-metallized glasses in region *5*.

In Figure 175 we present a triangular classification diagram, $\tau - \varrho - \alpha$. In the left-hand part of the triangle lie glasses that are mainly reflective, in the right-hand part those that are mainly absorptive.

Table 45 gives numerical information on several types of protective glasses.

Signal glasses

Rail, road, air, and water transportation systems have developed a range of coloured light signals. The chromaticity of signal glass depends both upon the light source used and upon the chromaticity of the coloured glass placed in front of the light source.

242

In order to simplify measurement and comparsion of signal glasses, standard sources (Planckian radiators) are prescribed for the following colour temperatures:

2 000 K   kerosene or oil lamp,
2 360 K   vacuum lamp or acetylene flame,
2 854 K   gas-filled lamp.

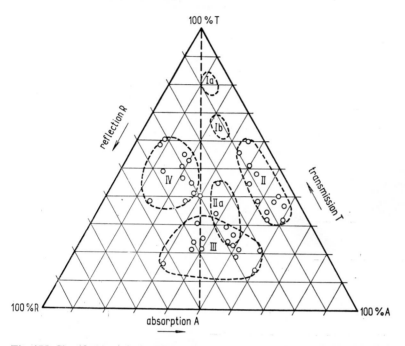

Fig. 175 Classification of protective glasses by predominantly reflective and predominantly absorptive properties: *Ia* — flat colourless glass; *Ib* — flat colourless glass in the form of a double pane; *IIa* — dethermal glasses coated with dielectric layers; *II* — dethermal glasses including Spectrofloat glass; *III* — protective glasses coated with metallic films; *IV* — glasses coated with dielectric layers.

Table 45

Glass	Thick-ness [mm]	τ_r	τ_{ef}	τ_s	ϱ_s	α_s
colourless window glass	5.5	0.82	0.85	0.90	0.08	0.02
dethermal	4.8	0.59	0.69	0.84	0.07	0.09
Reflex S	4.7	0.61	0.63	0.54	0.45	0.01
Reflex KP	5.2	0.53	0.58	0.57	0.38	0.05
Reflex dethermal	5.1	0.45	0.54	0.79	0.11	0.10
Spectrofloat Bronze	5.9	0.58	0.67	0.49	0.11	0.40

The regions of the individual colours of signal glasses in the chromaticity diagram xy are outlined by straight lines. The Czechoslovak standard CSN 01 2728 defines the coordinates of the intersections of these lines. The chromaticity of the signals must thus lie in the regions allocated to the various colours in the chromaticity diagram xy. In Figure 176 we present that diagram with the colour regions of signal glasses currently used in road transport (fixed light signals).

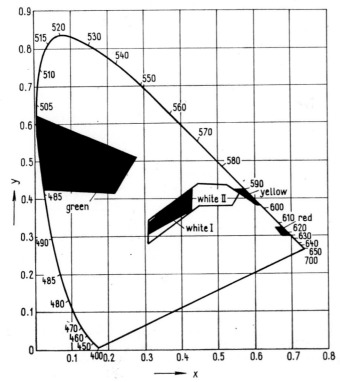

Fig. 176 Regions of colours of signalling fixtures in road transport (fixed signalling units).

In the measurement of the chromaticity of the signal light (filter), due account must be taken of the geometrical arrangement, viz. whether the element is a plane-parallel filter, a lens, or an element scattering light, etc. The measuring procedure is described in ref. [10].

Container glasses

Container glasses have recently been examined as far as their spectral transmittance is concerned. The ultra-violet and shortwave visible radiations are especially harmful to foods or medicines [14, 15]: among foods that can be particularly affected is milk (whose C, A, B_6 vitamins and riboflavin become neutralized), beer (whose quality is debased), eating oils (which decompose and become rancid), pharmaceuticals, etc.

244

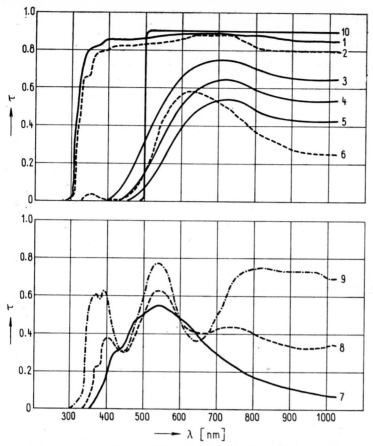

Fig. 177 Curves of special transmission in Czechoslovak container glasses: *1* — white glass (Kyjov);
2 — current container glass (Rudolfova huť); *3* — light brown bottle glass coloured with
Mn + Fe (Lesní Brána); *4* — brown bottle glass coloured with Mn + Fe (Nové Sedlo);
5 — dark brown bottle glass coloured with Mn + Fe (Lesní Brána); *6* — amber bottle glass
(Ústí nad Labem); *7* — olive-green bottle glass coloured with Fe + Mn (Nové Sedlo),
8 — emerald green glass coloured with Cr + Fe (Ústí nad Labem); *9* — emerald green glass for
dosing machines coloured with Cr + Fe (Nové Sedlo), *10* — the ideal curve of transmission
in container glass.

Container glass should therefore exhibit the highest possible capacity of absorbing the ultra-violet, the violet, and the blue part of the spectrum, and it should
manifest good transmittance in the region of the wavelengths longer than 500 nm.
From the curves of spectral transmittance (after [14, 15]) presented in Figure 177,
it follows that these requirements are satisfied best by brown bottle glass (coloured
with Mn + Fe), dark brown bottle glass (coloured with Mn + Fe), and amber
bottle glass.

It can thus be expected that the volume of brown and yellow glasses in the output
of container glasses will gradually increase.

5.3.2   Utility glasses, luxury glasses, and imitation jewellery glasses

Utility glasses

The use of coloured glasses in the manufacture of utility and luxury products substantially enhances both the artistic (aesthetic) value and the practical value of these products. They are usually made from coloured glass, but they can also be flashed and their surface can be painted, stained, or otherwise coated with chromatic layers.

The choice of the colour usually depends upon the aesthetic tastes of the artists and upon the schools of art they are associated with. One very important aspect is the expected price of the product.

Imitation jewellery glasses

Imitation jewellery glasses in the narrow sense of the word are simulated gems and pearls; in the broader sense of the word they are small glass products of various shapes. They are mostly produced from alkali-lime-silica glasses and from glasses containing lead monoxide. Glasses with a 30 wt. % to 50 wt. % content of lead monoxide are typical imitation jewellery glasses.

a)

b)

Fig. 178 The most current type of machine-cut stones with 17 facets.

Owing to the vagaries of fashion, the assortment must be varied and continually enlarged as far as the colour, hue, degree of opalescence, shape, cut, etc. are concerned. The requirement here is to ensure the reproducibility of the properties. From the point of view of the optical properties the products are evaluated in such a way that the tristimulus values are measured and chromaticity coordinates plotted in the chromaticity diagram, or so that they are calculated from the measurements of the spectral transmission of radiation.

The most common shape of the machine-cut gems are stones with seventeen facets (see Figure 178).

Most of the imitation jewellery products have silvered bottom facets in order that a light-reflecting effect may be obtained. The same effect is also achieved by vacuum coating these facets with a layer of highly reflective material, by gilding, iridizing, etc. The colour palette of the glasses from which simulated gems are produced is very rich, and the stones are usually referred to by the names of the genuine gem stones.

A unified system of classifying glasses used in the production of imitation jewellery has been introduced, alotting all raw glass materials different numbers. Eighteen numbers give information about the chromaticity and data on the products according to groups, types, design, and dimensions. This classification incorporates the four-digit classification which was in use until very recently, where the first figure denoted the colour, the second the degree of transparency or opalescence, and the overall appearance. The other two figures defined the intensity of the colour or the shades of the combination. For instance, for the designation of glass beads (rocailles) the first figure categories are:

0 — crystal	3 — blue	6 — bluish green
1 — brown	4 — grey	7 — pink
2 — violet	5 — green	8 — yellow
		9 — red;

and the categories of the second figure:

0 — clear	5 — agate
1 — opal	6 — moonlight
2 — alabaster	7 — flashed
3 — cloudy	8 — striped
4 — silken lustre	

Transparent colouration is manifested by glass that does not contain light-scattering particles, whereas opal, alabaster, and cloudy apply to opalescent (light-scattering) glass.

5.4 OPALESCENT GLASSES

Lighting glasses, sheet glasses, bent glasses, and hollow glasses, which modify the radiant flux so as to suit it to lighting purposes, are defined by various standards. For decorative lighting glasses these standards are optimal only.

Glasses for the lighting can be classified into groups, as shown in Table 46.

In slightly opalescent gasses transmittance is not measured unless it exceeds the value of 0.01. For highly opalescent glasses ($\tau < 0.01$) the so-called observable direct transmission must be determined by examining the specimen at a 40-cm distance and ascertaining whether the shape of the sharply defined 10 cm^2 surface of a 1 000 nt source placed 15 cm behind the specimen is fully discernible.

The lighting qualities of each type of glass will manifest themselves in a different way in sheet glass and in hollow glass. And this is why the lighting indices are given separately.

Evaluation of the lighting qualities of flat and bent glasses

The evaluation concerns

(a) total reflectance, ϱ_c, total transmittance, τ_c, and absorption, α_c,

by (b) luminosity and luminance curves of the reflected and the transmitted light (i.e. by indicatrix curves).

Table 46

According to scatter		Glasses	
transparent		colourless coloured	
diffusing	semi-transparent	matted figured	chemically mechanically regularly irregularly
	translucent	opalescent cloudy	massive flashed massive flashed

The distribution of the reflected and the transmitted light is visually represented by the reflectance and transmittance curves in polar or Cartesian coordinates. The luminance of the reflected or transmitted light for various directions of observation is also plotted with the help of indicatrix curves. From the luminance curves we can then determine the diffusion factor (CIE 1939) (see Section 4.4.3).

Table 47

Transmissive glasses			Lighting indices [%]				
			Light incident on	Transmission τ	Absorption α	Reflection ϱ	Diffusion ϑ
semi-transparent	matted		smooth surface	63—87	4—17	7—20	about 6
			matt surface	77—89	3—11	6—16	about 5
	figured		smooth surface	57—90	3—21	7—24	accord. to specimen
translucent	cloudy	opalescent	—	65—78	4—10	13—31	80
		highly opalescent	massive flashed	—	30—45	3—7	50—65	90
				—	45—60	2—5	35—50	80—90

In Table 47 are quoted the lighting indices for normal incidence of A light for a 2.8 mm thick glass.

248

Evaluation of the lighting qualities of hollow glass

Owing to multiple internal reflection, hollow glass manifests other values than flat glass of the same type, quality, and thickness. For a wholly enclosed sphere the transmission factor, τ_c, absorptance, α_c, and total efficiency, η_c, can be calculated from the values recorded on a flat glass specimen with the help of the equations

$$\tau_c = \frac{\tau}{1 - \varrho}, \tag{280}$$

$$\alpha_c = \frac{\alpha}{1 - \varrho} = 1 - \frac{\tau}{1 - \varrho}, \tag{281}$$

$$\eta_c = \frac{\tau}{1 - \varrho} = 1 - \frac{\alpha}{1 - \varrho}. \tag{282}$$

The surface luminance of a sphere is calculated from the relationship

$$B = \frac{\Phi \eta_c}{\pi^2 d^2}, \tag{283}$$

where B is the surface luminance of the sphere [nt],
$\quad \Phi \quad$ the radiant flux of the source [lm],
$\quad \eta_c \quad$ the efficiency of the sphere,
and $\quad d \quad$ the diameter of the sphere.

For an open sphere the solutions of the given relationships will yield only approximate results.

The efficiency of a sphere, η_c, is thus inversely proportional to α, proportional to τ, and again inversely proportional to ϱ. A hollow internally matted lighting fixture will thus be more efficient, and its scatter more perfect, than an externally matted lighting fitting.

With flashed opal glass the efficiency is higher and the scatter more perfect if the opalescent layer forms the inner wall of the lighting fixture.

Table 48

Glass	Luminous efficiency [%]
colourless	90
flashed white opal glass	80

For the quality of lighting glass the most decisive factor is its lighting efficiency, which is quoted for a sphere of 200 mm diameter with a 100 mm wide opening. The lowest values of lighting efficiency permitted are given in Table 48.

The method of measuring the luminous intensity of a lighting fixture is described in Section 4.4.5, and the luminous intensity curve is presented in Figure 158. For detailed information in the properties of lighting glasses and the modes of measurement the reader is referred to ref. [19].

5.5 GLASSES FOR CATHODE-RAY TELEVISION TUBES

The spectral transmission of light radiation is reduced in glasses from which the screens of television (TV) tubes, whether plain or colour, are produced, in order that a higher contrast may be obtained.

Table 49

Glass	Thickness [cm]	Total luminous transmission $\tau . 100$ [%]	Chromaticity coordinates for source C	
			x	y
Philips — CTV (Holland)	0.7	63.2	0.3064	0.3149
Schott — CTV (Mainz — GFR)	0.7	66.4	0.3039	0.3160
Sovirel — CTV (France)	0.7	66.3	0.3061	0.3151
Corning — CTV (USA)	0.7	75.4	0.3091	0.3153
Sovirel — BWTV (France)	1.0	44.4	0.3041	0.3150
USSR — BWTV	1.0	43.0	0.3065	0.3138
Romania — BWTV	1,0	41.0	0.3036	0.3144
AVTB (CSSR) — BWTV	0.7	57.7	0.3059	0.3159

CTV — colour TV
BWTV — black-and-white TV

In colour TV tubes it is further essential that the tint of the glass should not impair the colour scheme of the picture transmitted. The glasses from which colour or plain tubes are produced are therefore neutrally grey, so that the curves of spectral transmission within the range of the 400—700 nm waveband are approximately uniform without manifesting any pronounced absorption bands. As compared with colour TV tubes, plain TV tubes are slightly more absorptive in the quoted waveband.

250

The evaluation also includes the appraisal of chromaticity given by the chromaticity coordinates related to the CIE standard *C* white light.

In Table 49 the reader will find these characteristic properties quoted for several glasses.

References

[1] ČSN 71 0114. Optické sklo (Czechoslovak Standard No. 71 0114. Optical Glass). 1955.
[2] ČSN 70 8001. Křišťálová užitková skla (Czechoslovak Standard No. 70 8001. Crystal Utility Glasses). 1973.
[3] NF B 30-004. Verre Cristal, cristallin et verre sonore.
[4] B. S. 3828. Lead Crystal Glasses. 1964.
[5] 69/493/CEE. Directive du Conseil du 15. décembre 1969 concernant le raprochement des legislations des États membres relatives du verre cristal.
[6] Z 1997 A. Gesetz zur Kennzeichenung von Bleikristall und Kristallglas. GFR.
[7] S. Bachtík and V. Pospíchal, Zušlechťování skla (Refinement of Glass). SNTL, Prague, 1964.
[8] J. Wolf, Le travail et le faconnage decoratif de verre. L'Edition Universelle S. A., Bruxelles, 1942.
[9] ČSN 70 7103, Skleněné ochranné svářečské filtry (Protective Glass Welder's Filters). 1978.
[10] ČSN 01 2728. Barvy návěstních světel pro dopravu (Czechoslovak Standard No. 01 2728. Colours of Signal Lights for Transport). 1969.
[11] ČSN 70 5510. Světelně technické vlastnosti a hodnocení osvětlovacího skla (Czechoslovak Standard No. 70 5510. Lighting Properties and Evaluation of Lighting Glass). 1962.
[12] M. Kříž, Možnosti zlepšení ochranných vlastností skla proti sluneční radiaci (Possibilities of improving the protective properties of glass against solar radiation). Sklo ve stavebnictví, Ústí n. Labem, (1973) pp. 43—56.
[13] M. Dubček, Přímé měření průchodu sluneční energie sklem (Direct Measurement of the Passage of Solar Energy through Glass). Sklo ve stavebnictví, Ústí n. Labem, (1973) pp. 91—102.
[14] M. Kříž and J. Kadeřábek, Nové poznatky z oboru barvení obalových skel (Recent achievements in the colouration of container glasses). *Silikáty*, 7 (1963) pp. 215—230.
[15] M. Kříž, J. Kadeřábek, and V. Pacovský, Das Schutzvermögen des Verpackungsglases als Funktion seiner spektralen Durchlässigkeit. 7th International Congress on Glass, Brussels, 1965, III, 1—7.
[16] ČSN 70 7102. Skleněné optické části ochranných očních pomůcek pro průmyslové účely (Czechoslovak Standard No. 70 7102. Glass Optical Parts of Sight-Protecting Aids for Industrial Purposes). 1979.
[17] G. F. Brewster, Calculated X-ray mass absorption coefficients of glass components. *J. Am. Ceram. Soc.*, 35 (1952) (No. 8) pp. 194—197.
[18] J. Kocík, J. Nebřenský, and I. Fanderlik. Barvení skla (Colour Generation in Glass). SNTL, Prague, 1978.
[19] Coll. of authors, Osvětlovací sklo v interiéru (Lighting Glass in the Interior). SNTL, Prague, 1965.

SPECIAL APPLICATIONS OF THE OPTICAL PROPERTIES OF GLASSES

6.1   PHOTOSENSITIVE AND PHOTOFORM GLASSES

Photosensitive glasses are capable of retaining a latent (invisible) image if exposed to radiation of a certain wavelength. The latent image can be evoked and a visible picture obtained if the glasses are subjected to further thermal treatment.

The term *photoform* applies to the special photosensitive glasses in which the photosensitive metal (after exposure to radiation) will, in the course of thermal treatment, give rise to crystallizing nuclei of silicates or fluorides. It is a particular property of these glasses that they are capable of creating a plastic image of the copied original owing to the fact that the crystallized mass (irradiated and photonucleated) will dissolve at a rate different from the rate of dissolution of the original glass [1, 2].

The basic substance of photosensitive glasses is the presence of so-called photosensitive elements, gold, silver, copper, etc., which are present in small quantities. Under an activizing radiation, mostly ultra-violet, a quantum hv_1 will free an electron which, reacting with the ion of the photosensitive metal, will give rise to an electroneutral atom and liberate one quantum of radiation hv obeying the reaction (284), viz.

$$Ag^+ + hv_1 \nearrow (Ag^+)^* \\ \searrow Ag^{2+} + e^-. \tag{284}$$

At points that have been exposed to radiation active centres of $(Ag^+)^*$ will form and simultaneously a second reaction will liberate electrons e^- to form neutral atoms of Ag° during the ensuing thermal treatment, viz.

$$(Ag^+)^* + e^- \rightarrow Ag^\circ + hv + *. \tag{285}$$

These neutral atoms give rise to crystal nuclei which grow to colloidal dimensions. These particles of colloidal dimensions then give rise to the colouration and crystalline opalescence of the places irradiated. In contrast to solarization or photochromic reactions, the colour centres thus formed are stable at normal temperatures.

The activation of the photosensitive elements is favourably affected by Ce^{3+}, so that according to the reaction

$$Ce^{3+} + hv_1 \rightarrow Ce^{4+} + e^- \tag{286}$$

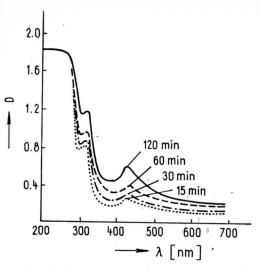

Fig. 179 Spectral characteristics of photosensitive glass of the following composition: Na₂O — 15 wt. %, ZnO — 10 wt. %, SiO₂ — 75 wt. % + Ag 0.07, CeO₂ 0.05, and SnO₂ 0.2 (g) per 100 g of glass. Thermally treated for 30 min at 530 °C. Effect of the duration of irradiation.

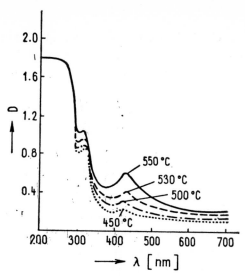

Fig. 180 Spectral characteristics of photosensitive glass of the composition quoted in Figure 179. Irradiated for 60 min, thermally treated for 30 min. Effect of the temperature of thermal treatment.

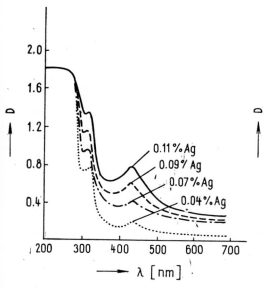

Fig. 181 Spectral characteristics of photosensitive glass of the composition quoted in Figure 179. Irradiated for 60 min, thermally treated for 30 min at 530 °C. Effect of the concentration of Ag.

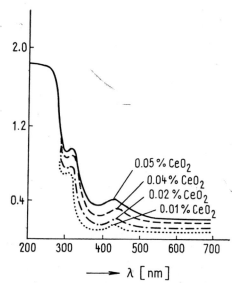

Fig. 182 Spectral characteristics of photosensitive glass of the composition quoted in Figure 179. Irradiated for 60 min, thermally treated for 30 min at 530 °C. Effect of the concentration of CeO₂.

253

Ce^{3+} will oxidize to Ce^{4+} and liberate an electron. The freed electrons will thus favourably affect the activation of the photosensitive metals and facilitate their reduction. A disadvantage, however, is that Ce^{3+} absorbs the activating radiation, so that the image can form in the surface of the glass only.

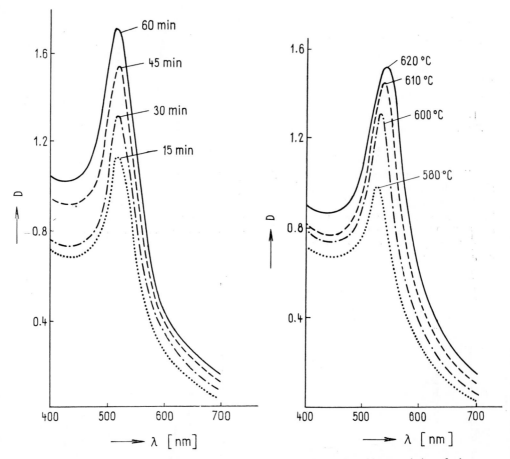

Fig. 183 Spectral characteristics of photosensitive glass of the following composition: Na$_2$O — 11 wt. %, Al$_2$O$_3$ — 2 wt. %, BaO — 9 wt. %, CaO — 6 wt. %, SiO$_2$ — — 72 wt. % + Au 0.04, CeO$_2$ 0.05, and SnO$_2$ 0.02 (g) per 100 g of glass. Thermally treated for 60 min at 600 °C. Effect of the duration of irradiation.

Fig. 184 Spectral characteristics of photosensitive glass of the composition quoted in Figure 183. Irradiated for 30 min, thermally treated for 60 min. Effect of thermal treatment.

Reducing substances, e.g., tin or antimony, facilitate the growth of crystal nuclei into larger particles of colloidal dimensions. Inhibitors, on the other hand, neutralize the photosensitivity of metals: e.g., ions absorptive in the ultra-violet region of the spectrum (thus absorbing activating radiation), and those preventing the formation of the active centres in photosensitive metals. This applies particularly to iron trioxide, titanium dioxide, arsenic trioxide etc.

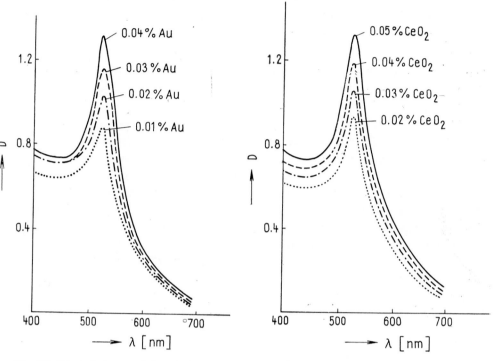

Fig. 185 Spectral characteristics of photosensitive glass of the composition quoted in Figure 183. Irradiated for 30 min, thermally treated for 60 min at 600 °C. Effect of the concentration of Au.

Fig. 186 Spectral characteristics of photosensitive glass of the composition quoted in Figure 183. Irradiated for 30 min, thermally treated for 60 min at 600 °C. Effect of the concentration of CeO_2.

Dwivedi and Nath [20, 21] studied the kinetics of the photochemical reactions in glasses containing silver and gold at various times of exposure and of thermal treatment, and for various concentrations of silver, gold, and ceric oxide. The results of the measurement of the spectral characteristics are presented in Figures 179 – 186.

The results of the vork of these authors point to the following sequences of reactions:

$$Ce^{3+} + h\nu \rightarrow Ce^{4+} + e^-,$$
$$Ag^+ + h\nu \rightarrow (Ag^+)^*,$$
$$\rightarrow Ag^{++} + e^-,$$
$$Ag^+ + e^- \rightarrow Ag^0, \tag{287}$$

255

$$Ag^{++} + 2e^- \rightarrow Ag^0,$$
$$x(Ag^0) \rightarrow (Ag^0)x,$$

and

$$Ce^{3+} + h\nu \rightarrow Ce^{4+} + e^-,$$
$$Au^+ + e^- \rightarrow Au^0, \qquad\qquad (288)$$
$$x(Au^0) \rightarrow (Au^0)x.$$

With silver present, the dimensions of the Ag^0 particles range approximately between 10 nm to 20 nm (with an absorption band at 405 nm); the dimensions of the Au^0 particles between approximately 5 nm to 60 nm (with an absorption band at 530 nm).

If the nucleus of a metal resulting from a photoreaction is large enough (~ 8 nm), the nucleation of the crystallization of silicates or fluorides may occur in glasses of the appropriate composition. Frequent use is made of crystallization within the systems $Li_2O \cdot SiO_2$, $Li_2O \cdot 2\,SiO_2$, and $BaO \cdot 2\,SiO_2$. In dilute hydrofluoric acid the crystalline phase will dissolve about 30 times $-$ 60 times as quickly as in the basic glass. Only the velocity of dissolution of $BaO \cdot 2\,SiO_2$ is lower, and this fact is made use of in the manufacture of *embossed glass*. Such glasses are generally referred to as photoform.

In *photoceramic glass* use is made of the fact that crystallization, which is photonucleated in order that a fine crystal texture may be achieved, will give rise to lithium aluminosilicates, which, at the temperature of about 700 °C to 900 °C, will decompose to form β-quartz and other crystalline phases.

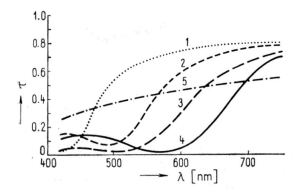

Fig. 187 Spectral characteristics of photosensitive glasses:
1 — yellow; *2* — orange; *3* — red; *4* — purple; *5* — grey.

As already mentioned above, photosensitive glasses contain photosensitive metals — gold, silver, copper etc. Gold is used in concentrations of 0.001 to 0.05 %, silver in the form of silver chloride in 0.0001 to 0.3 % concentrations, and copper in the form of cupric oxide in concentrations of 0.05 to 1.0 %. Palladium can only be used in combination with gold or silver, because palladium itself is not photosensitive; it changes the original hue to brown or grey. The quantity usually applied is 0.001 % to 0.02 %.

Photosensitive glasses which contain gold are red, purple, violet, and blue. These colours depend upon the composition of the parent glass and upon the concentration of the gold. For instance, glasses containing silver are yellow to amber, and copper gives red colour. Examples of the composition of photosensitive and photoform glasses are given in Table 50.

Figure 187 gives the curves of spectral transmittance of photosensitive glasses.

6.2   PHOTOCHROMIC GLASSES

By photochromic we refer to glasses in which colour is generated, viz. colour centres are formed, while they are exposed to radiation of certain energies (wavelengths). If the exposure to activating radiation is interrupted, these glasses will lose colour, viz. the colour centres will decompose. In contrast to the common types of coloured glasses, the colour centres of which are stable, the colour centres of photochromic glasses are metastable.

Photochemically sensitive and highly applicable in this respect are the silver halides, copper and cadmium halides (the latter being more economical), europium, and cerium. From among the other, less frequently applied, systems mention should be made of, for example, glasses containing chromium, molybdenum, tungsten, glasses of the Hackmanite type of compostition, etc. [43].

The most advantageous from the point of view of their properties, and, consequently, from the point of view of practical applicability, are considered to be the photochromic glasses containing the halides of silver. We shall therefore deal with these glasses in more detail.

Photochromic glasses containing silver halides

In photochromic glasses containing silver halides the activating radiation hv_1 (ultra-violet to short-wave visible) gives rise to colour centres. As soon as exposure to radiation is interrupted, photons of energy hv_2 (long-wave visible to thermal radiation) will destroy these centres [3, 4]. The latter are formed according to the rather simplified reaction, viz.

$$AgX \underset{hv_2}{\overset{hv_1}{\rightleftharpoons}} Ag° + X°, \qquad (289)$$

where X stands for chlorine, bromine.

This photochemical reaction takes place in the separated phase of silver halides, which is then present in the final photochromic glass in crystalline form.

The reserved photochemical reaction can take place provided the products of the reaction do not react with the other substances present, and do not diffusively leave the region of the reaction and grow into stable aggregates, as is the case in photographic layer, where the reaction can be rewritten as

$$n\,AgX \cdot \overset{hv_1}{\longrightarrow} n\,Ag° + n\,X° \qquad (290)$$
$$\downarrow$$
$$(Ag°)\,n$$

Table 50

Component	Weight %									
	Photosensitive			Photoplastic			Photochromic			Polychromatic
	1	2	3	1	2	3	1	2	3	
SiO_2	70.6	66.8	70.6	85.0	81.0	58.0	59.1	59.1	59.1	69
Li_2O				10.0	12.5					
K_2O	17.4	23.7	12.8		2.5	15.0	10.8	10.8	10.7	15.8
Na_2O	1.5		1.9		4.0		9.3	9.3	9.3	6.8
Al_2O_3	8.4		5.9	5.0						
CaO			8.8							
BaO						27.0				
Ag_2S		0.14								
ZnO		9.4	0.02							
Sb_2O_3										4.8
$AgCl$	0.04		0.04	0.08	0.02	0.08				0.2
CeO_2			0.01	0.02	0.025	0.002				
Au										
B_2O_3							20.0	20.0	20.0	2.3
F							1.5	1.5	1.5	0.01
Ag							0.47	0.42	0.47	
Cl							0.30		0.2	
CuO	0.1						0.015	0.015	0.015	1.0
Br								0.72	0.5	0.05
SnO_2	0.25									
NH_4Cl	1.8									0.05
Ce^{3+}										0.05

In order that sensitivity to activating radiation may be increased, photochromic glasses are doped with *senzitizers*. One of the most effective of these agents is copper protoxide used in commercially produced photochromic glasses. Reaction 289) can thus be rewritten as

$$Ag^+ + Cu^+ \xrightleftharpoons[hv_2]{hv_1} Ag^\circ + Cu^{2+}. \tag{291}$$

The kinetics of formation and destruction of the metastable colour centres are also influenced by the dimension of the separated phase of silver halides. The phase separation of the silver halides is caused by the subsequent thermal treatment

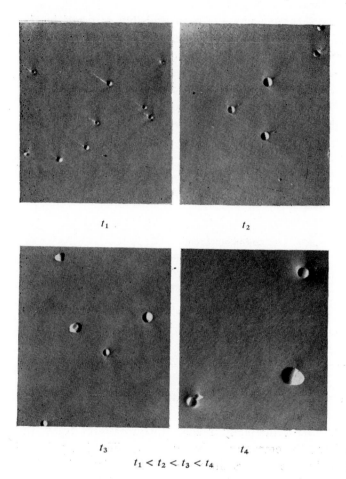

t_1

t_2

t_3

t_4

$t_1 < t_2 < t_3 < t_4$

Fig. 188 Electron micrographs of the separated phase of silver halides. Increasing temperature associated with the dimension of the separated phase.

to which the glass melt is subjected. The separation is most probably effected by the mechanism of nucleation and growth [5]. It is limited by the uppper critical temperature, T_c, the lower limit being set by the cessation of diffusion due to the increased viscosity of the glass. From [6] it follows that a low temperature thermal treatment ($350° - 575$ °C) will produce colloidal particles of Ag°. It is assumed that the Ag° colloidal particles first nucleate the separation of basic glass and that the local oversaturation of the peripheral regions of the separated phases by the Ag^+ and X^- ions leads to the precipitation of AgX. It is further assumed that Ag^+ and X^- dissolve more readily in the separated phase of the basic glass than in the surrounding phase.

At an advanced stage of separation of the silver halides (rising temperature and prolonged thermal treatment) the dimension of the separated phase will also grow. The number of separated regions in the specific volume of the basic glass will, however, simultaneously decrease (see Figure 188).

This rearrangement of the separated phase is most probably governed by the mechanism of diffusion. If the assumption may be made that with the increase of regions of the separated phase of silver halides the system will reduce its interfaces by enlarging its separated regions, this process could be interpreted as the so-called Ostwaldian process of maturation. In this case the separated regions, the dimensions of which have not reached the critical radius, r^*, will have a tendency to dissolve and to diffuse the ions towards the larger and more stable regions.

Table 51

t [°C]	Dimension [nm]	Description
550	26	non-photochromic
575	31	slightly photochromic
600	36	slightly photochromic
625	50	photochromic
650	68	photochromic
675	76	photochromic
700	100	photochromic
750	150	photochromic

As soon as the critical temperature, T_c, has been attained, the separated phase of halides will dissolve in the surrounding phase.

Cooling glasses that have been thermally treated at temperatures below the critical temperature, T_c, down to temperatures below the solidification point of AgX will cause the separated phase to crystallize, and if the desired dimensions have been reached, the glass will acquire photochromic properties.

Table 51 contains data on the dependence of the dimensions of the separated phase of silver chloride upon the temperature of thermal treatment. The dimensions of the separated phase have been determined by measurement of the angular distribution of the scattered radiation.

The time-related variations in the concentration of the absorbing, metastable colour centres in the course of their formation and destruction can (after [7]) be expressed by the linear differential equation

$$\frac{dC}{dt} = k_d \Phi_d A^* - (k_f \Phi_f + k_d \Phi_d + k_t) C, \tag{292}$$

where C denotes the concentration of the colour centres,

t		the time,
k_d		the velocity constant of the formation of the colour centres,
k_f		the velocity constant of the optical destruction of the colour centres,
k_t		the velocity constant of the thermal destruction of the colour centres,
Φ_d		the intensity of the radiant energy activating the formation of the colour centres,
Φ_f		the intensity of the radiant energy activating the destruction of the colour centres,
and	A^* is	the concentration of the active centres.

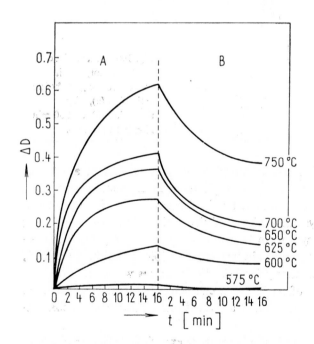

Fig. 189 Kinetics of the formation (A) and destruction (B) of the colour centres, ΔD denoting change in optical density.

From equation (292) it follows that the change in the concentration of the metastable colour centres depends, from an energy point of view, upon which of the two processes will prevail, the process of activation or the process of destruction. The equilibrium of these processes is thus of a dynamic character, and as the activated state involves a higher energy level, the termination of the activation will start processes of relaxation tending towards a transition to the energetically less intensive level of the original state. After Aranjo [43] and Fanderlik [4] the darkening and bleaching mechanisms of photochromic glasses are believed to be quite analogous to the photolytic effects found in bulk silver halide crystals. Figure 189 illustrates the kinetic characteristics of the formation and destruction of the colour centres in glasses of various dimensions of the separated phase of silver chloride (see Table 51). From the relationships seen in Figure 189 it can be inferred that with the dimension of the separated and, after cooling, of the crystalline phase of silver chloride growing, the process of formation of the colour centres will become more rapid after activation, and the concentration of the metastable colour centres will also be increased.

On the other hand, the growth of the size of the separated phase will be associated with deceleration of the destruction of the metastable colour centres.

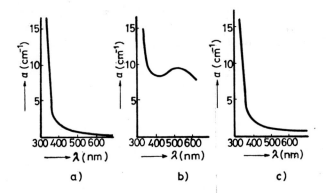

Fig. 190 Characteristic spectral absorption in photochromic glass before activation (a), after activation (b), and 30 min after activation (c).

The fact that the crystalline phase of silver chloride is located in a vitreous matrix renders the movement of the products of photochemical reaction impossible, or it may be confined to distances limited by the dimensions of these products. With the dimension of the crystalline phase of silver chloride increasing during activation, the colloidal silver will form larger, more stable aggregates (n Ag$^\circ \to$ $\to$ (Ag$^\circ$) n), so that although increased absorption of visible radiation is attained, the process of destruction of the newly formed colour centres will be retarded.

The activated colour centres of Ag$^\circ$ have a characteristic spectral absorption. Figure 190 gives the absorption spectrum of activated photochromic glass containing silver chloride.

It follows from theoretical research and laboratory experiments that the kinetics of the photochemical reactions taking place in photochromic glasses are very similar to the analogous processes in crystals of silver halides. The differences are caused mostly by the dimensions of the crystals and the medium in which they are located.

Photochromic glasses find wide application in the production of eye-protecting sun glasses. They have also been made use of in the image-forming equipment for analogue information, in holographic recording, and further applications seem most likely, e.g., in memory equipment, integrating diaphragms, etc. Table 50 presents information on the composition of photochromic glasses.

6.3   POLYCHROMATIC GLASSES

The most recent type of glass containing photosensitive substances is polychromatic glass (e.g., photosensitive, photoform, and photochromic glasses) developed by Corning Glass Works [41, 27].

At room temperature the ultra-violet radiation will first make the glass form a latent image. The exposure is to 300 nm wavelength radiation for a period of five seconds to five minutes. The reaction thus started can be rewritten as

$$Ag^+ + Ce^{3+} \xrightarrow{h\nu} (Ag^\circ)^* + Ce^{4+}, \tag{293}$$

and the colour of the glass will depend upon the period of the exposure to radiation, as shown in Figure 191.

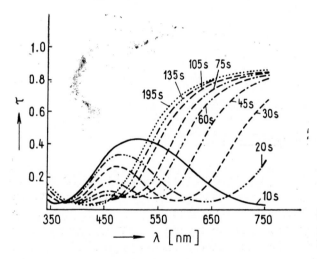

Fig. 191 Spectral characteristics of polychromatic glasses for various times of first exposure.

A first treatment at $450° - 500 °C$ will produce sub-colloidal silver nuclei, temperatures between $500 °C - 550 °C$ will then give rise to nucleated pyramid crystals NaF (Na, Ag) X (see Figure 192), viz.

$$n\,Ag^\circ \xrightarrow{\sim 450 °C} (Ag^\circ)\,n, \tag{294}$$

$$(Ag^\circ)n + XR^+ + xX^- \xrightarrow{\sim 520 °C} C \triangleright.$$

The final exposure to radiation of wavelength 300 nm (for ten minutes to one hour) will produce latent silver in the peak of the crystals,

263

$$C \triangleright \frac{h\nu}{300 \text{ nm}} C' \triangleright , \hspace{4cm} (295)$$

and the ensuing thermal treatment at 300 °C−410 °C a typical colour image, as shown in Figure 193 (from publicity material of Corning Glass Works Co.).
In equations (293) to (295)

R⁺ stands for Na⁺, Ag⁺,

R^+ stands for Na^+, Ag^+,
X^- for F^-, Cl^-, Br^-, (J^-),
$C \triangleright$ for pyramidal crystal NaF (Na, Ag) X, and
$C' \triangleright$ for the coloured pyramidal crystal with anisotropic silver in the peak.

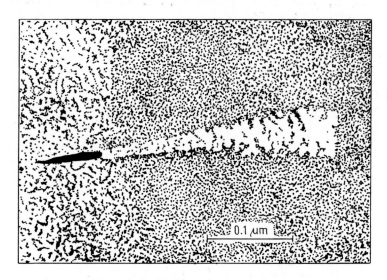

0.1 μm

Fig. 192 Typical pyramidal crystal in polychromatic glass.

The dimensions of the crystals range between 0.3 μm and 1 μm. The silver in the peaks of the crystals is likely to have been produced by fractional crystallization during the cooling of solid solutions of NaBr − AgBr.

The colouration is caused by the selective absorption of light by the anisotropic particles of silver in the peaks of the pyramidal crystals of complex halides, and it depends upon the shape of the silver particles. Changes in the colouration can be obtained ranging from green, via blue, violet, red, orange, to yellow, which corresponds to spherical colloidal particles in glasses.

6.4   SOLARIZATION OF GLASSES

Another kind of photochemical reaction taking place in glasses containing photochemically sensitive substances after they have been exposed to radiation is solarization. The reaction is activated by the energetically intensive ultra-violet component of solar radiation, the resulting effect being the generation of undesired colouration [8].

This type of photochemical reaction was first observed in glasses decolourized by manganese:

$$4MnO + As_2O_5 \underset{h\nu_2}{\overset{h\nu_1}{\rightleftarrows}} 2\,Mn_2O_3 + As_2O_3. \tag{296}$$

If iron oxide is present, and the glass has been refined by arsenic, solarization will start the reaction

$$4FeO + As_2O_5 \underset{h\nu_2}{\overset{h\nu_1}{\rightleftarrows}} 2\,Fe_2O_3 + As_2O_3. \tag{297}$$

Simultaneous occurrence of Mn^{3+} and Fe^{3+} ions will further increase the intensity of the post-irradiation colouration.

Owing to the fact that these photochemical reactions are associated with meta-stable colour centres, thermal energy can destroy the colour centres arising in this way. Reversible reaction will bleach the colour.

In refining and decolourizing glasses by CeO_2 in the presence of minute quantities of arsenic (or titanium dioxide) we get the reaction

$$2\,Ce_2O_3 + As_2O_5 \underset{h\nu_2}{\overset{h\nu_1}{\rightleftarrows}} 4\,CeO_2 + As_2O_3, \tag{298}$$

and the glass will acquire a yellow to grey tint.

The effect of solarization can be utilized, e.g., in measuring the amount of solar radiation. In glasses containing vanadium pentoxide and cerium oxide the reaction is

$$V_2O_3 + Ce_2O_3 \underset{h\nu_2}{\overset{h\nu_1}{\rightleftarrows}} 2\,VO + 2\,CeO_2, \tag{299}$$

and the glass becomes light grey to red.

The presence of lead monoxide stabilizes the resistance of glasses to solarization in the same way as the exposure to γ-radiation.

In accordance with present knowledge about the effect of solarization, for example, arsenic trioxide is not applied in refining commercially produced crystal glasses that do not contain either lead or barium monoxide; the aim is as low an iron trioxide content as possible.

6.5   GLASSES STABILIZED AGAINST THE EFFECT OF RADIATION

From among the various types of energy-intensive radiation that can strike glasses in particular applications the most frequent are photons (γ-radiation, X-ray radiation), electrons (β-radiation), charged atomic particles, and neutrons.

If γ-radiation of adequate energy interacts with glass, *Compton scatter* will give rise to ionization and to the *photoelectric effect*. β-radiation also ionizes atoms. Charged atomic particles (α-particles, protons, fission debris) have a similar effect. The resulting effects of the proton, the electron, and the charged atomic particles are thus analogous, the difference lies only in their penetrating capacities. Photons penetrate glass to a fairly considerable depth despite the low intensity of their effect. Electrons, on the other hand, exhibit a lower penetrating power, yet their effect is more intensive. The effect of atomic particles is particularly intensive, the depth of penetration is, however, very shallow; and neutrons affect the nuclei of the atoms.

Photons (γ-radiation) and electrons (β-radiation) give rise to colour centres by releasing electrons which, in the simplest case, start chemical processes of the oxidation-reduction type [9]. This is particularly the case if multivalent ions are present in the glass. The so-called shifting processes can also occur, displacing the atoms, or ions, from their equilibrium positions to interstitial positions; the displacement involves an energy change in the state of the electrons. After the anion has left its position, the vacancy will act upon the electrons as if it were positively charged. The moving electron, which has been released by incident radiation, will easily be trapped, and a new centre with typical absorption will come into existence. If this absorption is within the visible region of the spectrum, he centre will be a colour centre [9, 19, 44].

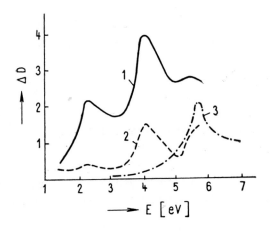

Fig. 194 Characteristic absorption bands in irradiated quartz glasses: 1 — glasses of the first group (Ultrasil, Infrasil); 2 — glasses of the second group (Homosil, Amersil); 3 — glasses of the third group (Corning 7940, Spectrosil, Suprasil).

High-purity quartz glass is resistant to γ-radiation, and no colour centres therefore appear. Irradiation of the other types of quartz glasses, however, will produce colour centres. Three absorption bands, viz. at 210, 300, and 550 nm are particularly intensive, as shown in Figure 194 (after Levy [28] and Cohen [29]).

The 215 nm absorption band is present in all irradiated quartz glasses, no matter how they have been produced. In glasses resistant to radiation, viz. very pure quartz glasses, however, 300 nm and 550 nm absorption bands either do not appear at all, or if they do, their intensity will be minimal.

Classifying quartz glasses from the point of view of the irradiation and the presence of impurities, we get (after Byurganovskaya, Vargin, Leko, and Orlov [19]) three groups:

(a) glasses with a high concentration of impurities ($10^{-3} - 10^{-2}$ wt. %) melted in contact with graphite (curve 1 in Figure 194);

(b) glasses melted from natural materials or from synthetic $SiCl_4$ in the flame of an oxy-hydrogen blowpipe. Depending upon the purity of the raw materials and the method of melting, the range of the variations of the structure of the absorption spectrum and the intensity of the absorption bands is much larger. The characteristic feature is the low intensity of the 550 nm absorption band at low irradiations. This band starts forming at an irradiation of 258 [$C . kg^{-1}$] onwards;

(c) glasses melted from high purity raw materials or those subjected to supplementary treatment by special technologies. The absorption spectrum consists of the 210 nm $-$ 215 nm absorption band, and of fairly weak bands at 240 nm to 260 nm. The absorption spectra of the glasses in the region of the wavelengths shorter than 220 nm (measured in vacuo) are presented in Figure 195 (after Nelson and Weeks [30]).

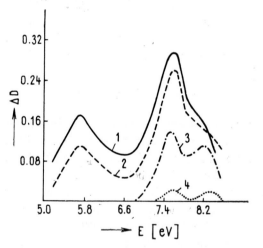

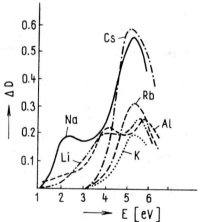

Fig. 195 Absorption spectra of quartz glasses in the ultra-violet region: 1 — glass exposed to γ-radiation of [10^5 C . kg^{-1}]; 2 — irradiated glass heated to 250 °C for 30 min; 3 — irradiated glass further heated to 450 °C for 30 min; 4 — irradiated glass further heated to 625 °C for 30 min.

Fig. 196 Effect of Na, Li, Al, K, Rb, and Cs on the character of the absorption spectra in quartz glass: Na — 0.2 wt. %, K, Rb, Cs, Li — 0.5 wt. %, Al — 2 mol. %. Exposed to γ — radiation of 2580 [C.kg^{-1}] (Li and Al 258 [C.kg^{-1}]).

Lell [31, 32] studied the effect of sodium, lithium, aluminium, potassium, rubidium, and caesium upon the character of the absorption spectra of quartz glass exposed to γ-radiation. The results are presented in Figure 196.

Very similar to quartz glass are silicate glasses, which, having a low alkali content, manifest almost the same absorption characteristics. Building metallic ions into the structure of the glass will "extend" the structure owing to the formation of non-bridging oxygens. The electron transitions from the non-bridging oxygens into the conductance band then require far less energy than the transitions from the bridging oxygens, so that the absorption is shifted towards longer wavelengths. Boric, phosphoric, and germanic glasses show similar characteristics.

For irradiated coloured glasses the theoretical explications are rather complex, because these glasses already contain colour centres of typical spectral absorption before they have been exposed to radiation. Irradiation can further make the colouring components undergo reduction changes, and some of these components can even act as stabilizers countering the effects of radiation. Only aggregate effects can therefore be observed.

Exposure to $0.258 \, [C . kg^{-1}]$ of γ-radiation will not produce any recordable changes.

Irradiation by 0.258 to $2.58 \, [C . kg^{-1}]$ of γ-radiation will give rise to colour centres, and glasses must be stabilized.

γ-radiation exceeding $2.58 \, [C . kg^{-1}]$ will trigger off a rapid development of colour centres. Equilibrium concentration is reached at $25\,800 \, [C . kg^{-1}]$, so that higher irradiation will not produce any further change in the concentration of the colour centres. The structure of the glass can, however, also be destroyed. The detailed nature of the radiation-induced defect centres in glasses has been reported by Friebele an Griscom [44].

Stabilization of glasses

The introduction of certain substances into glass will raise the resistance of that glass to the effects of radiation. These substances are most often cerium, antimony, arsenic, iron, cobalt, manganese, nickel, vanadium, copper and others. Of course, their stabilizing capacities are very far from being equal.

The presence of these substances in glass will curb absorption produced by irradiation in the visible region of the spectrum; they can, however, give rise to their characteristic absorption in other regions of the spectrum.

Two aspects of their behaviour must thus be considered:

(a) they counteract the radiation energy, and
(b) they counteract the electrons.

For instance, the change in colour of a glass containing V^{3+} due to the effect of radiation can be expressed as

$$V^{3+} \, (green) + e^- \rightleftarrows V^{2+} (pink). \tag{300}$$

On the other hand, for example, in glass with a Ce^{4+} content the change in valency due to irradiation will not be followed by a change in the colour of the glass, viz.

$$Ce^{4+} + e^- \rightleftarrows Ce^{3+}. \tag{301}$$

A number of authors suppose the protective effect of cerium in glass to be based upon reaction (301), whereas other authors assume the reaction to be taking the opposite direction.

Figure 197 gives curves of spectral transmission in the BK 7 optical glass, non-stabilized and non-irradiated (*1*), non-stabilized and activated (irradiated) by 25 800 $[C \cdot kg^{-1}]$ (*2*), stabilized by 2.5 wt.% of ceric oxide and non-irradiated (*3*), stabilized by 2.5 wt. % of ceric oxide and irradiated by 25 800 $[C \cdot kg^{-1}]$ (*4*), stabilized by 1.5 wt. % of antimony trioxide and irradiated by 25 800 $[C \cdot kg^{-1}]$ (*5*), stabilized by 1.5 wt % of ceric oxide plus 1.5 wt. % or antimony trioxide and irradiated by 25 800 $[C \cdot kg^{-1}]$ (*6*).

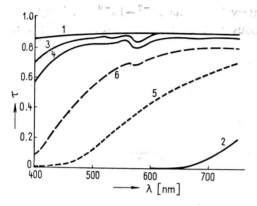

Fig. 197 Spectral characteristics of BK 7 optical glass: *1* — non-stabilized and not exposed to radiation; *2* — non-stabilized and irradiated; *3* — stabilized by 2.5 wt. % of CeO_2 and not exposed to radiation; *4* — stabilized by 2.5 wt. % of CeO_2 and exposed to radiation; *5* — stabilized by 1.5 wt. % of Sb_2O_3 and irradiated; *6* — stabilized by 1.5 wt. % of CeO_2 + 1.5 wt. % of Sb_2O_3 and irradiated. Irradiated by 25 800 $C.kg^{-1}$.

Glasses with an increased content of lead monoxide manifest a higher resistance to radiation, thus the addition of ceric oxide need not have a more pronounced effect. As far as the stabilizing capacities of antimony and arsenic trioxide are concerned, their increased concentration makes the absorption edge shift towards the shorter wavelengths, so their spectral transmission is improved before they are exposed to radiation. In the presence of ceric oxide, however, they block the (stabilizing) effect of cerium, which results in an overall weakening of its protective capability.

The process of formation of colour centres by irradiation is reversible, being, however, rather slow without the intervention of external energy. The colour centres can be decomposed by heating the glass, exposing it to solar energy or to the radiation of a luminescent lamp. The speed of the reversible process depends upon the mobility of the electrons in the glass, which is conditioned by the structure of the glass as far as its chemical composition and thermal history are concerned.

6.6   LUMINESCENCE OF GLASSES

According to the method of excitation and the duration of luminescence after excitation has ended, luminescent processes can be classified into several groups.

We need to distinguish between *photoluminescent* and *thermoluminescent* (or radioluminescent) glasses and also between *fluorescence* (luminescence disappears after excitation has ended), and *phosphorescence* (luminescence lasts for a certain period after excitation has ended).

For fluorescence to appear, it is essential that a certain amount of energy should be absorbed in the glass. According to Stokes the frequency of the exciting radiation that causes fluorescence is higher than that of the fluorescent radiation itself. It follows from Figure 198 that luminescence appears in glasses containing so-called luminescent centres [10].

Between the excitation process (absorption of radiation) and the emission of the absorbed energy there is a short interval of some $10^{-7}-10^{-9}$ sec. They can thus in practice be viewed as simultaneous, and the term applied is fluorescence.

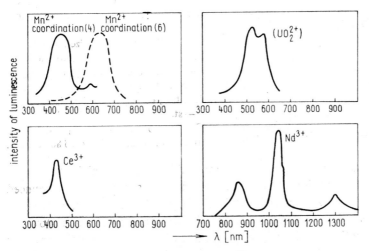

Fig. 198 Curves of luminescence in glasses doped by Mn^{2+}, Ce^{3+}, UO_2^{2+}, Nd^{3+}.

If, however, during the excitation an electron gets into an orbit from which transition to the ground state is forbidden, we have the so-called unstable state, the duration of which is limited only by collisions with other atoms. When excitation ends, emission (viz luminescence) sets in. This process is termed phosphorescence.

Figures 199 and 200 (after Peyches [11]) give a schematic representation of the course of photoluminescence and thermoluminescence.

As far as attenuation of fluorescence is concerned, we can claim that in almost all cases the intensity of fluorescence will decrease if the temperature rises, and vice versa. Raising the temperature to 300 °C−400 °C will cause fluorescence to vanish in the majority of glasses, the emission bands will shift towards longer wavelengths and fine structure will disappear. For instance, owing to this process

270

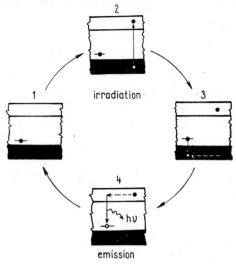

Fig. 199 Schematic diagram of the process of photoluminescence.

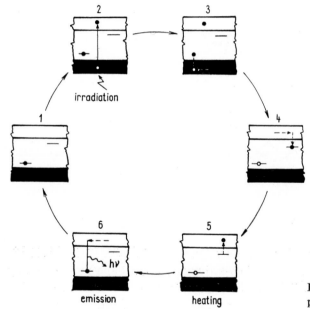

Fig. 200 Schematic diagram of the process of thermoluminescence.

the colour of a glass containing uranium will change from bluish green to yellowish green.

Concentration quenching, an effect of decreasing fluorescence, probably arises if the number of fluorescent centres exceeds a critical value. The interaction of fluorescence centres with atoms, ions, or molecules present in the glass (e.g., Fe^{2+}, Fe^{3+}, etc.) can then give rise to attenuation of fluorescence.

If the fluorescence centres are crystalline, e.g., SnS, CaFe, NaF, KF, $3 Ca_3 (PO_4)_2 . CaF_2$, $ZnSiO_3$, etc., the influence of composition of the glass on fluorescence is very far from being significant.

Glasses containing energy-isolated atoms or molecules are those with, for example, a silver, or cadmium sulphide, content; and glasses containing fluorescent ions can be produced by adding the rare earths, manganese, uranium, copper, thorium, tin, lead, vanadium, etc. These two latter groups include the glasses whose fluorescence centres are more or less influenced by the composition of the glass.

6.7 STIMULATED EMISSION IN GLASSES – QUANTUM GENERATORS OF LIGHT

Glass lasers are increasingly being used as sources of coherent mono-chromatic radiation, and have an important place along with other types of lasers.

A very simplified schematic diagram of a laser with a glass resonator can be seen in Figure 201.

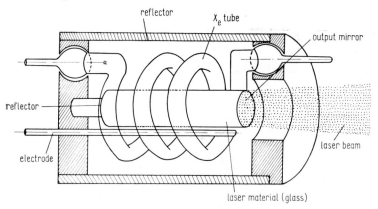

Fig. 201 Schematic diagram of a laser.

Many glasses have been doped with the trivalent ions, Nd^{3+} Yb^{3+}, Ho^{3+}, Er^{3+}, and others, pure or with a number of co-activators. One of the most currently used glass lasers contains Nd^{3+} ions. The absorption spectrum and the emission spectrum are determined by the scheme of energy levels of Nd^{3+}, particularly of its three electrons in the incomplete $4f$ level, shielded by the higher $5s$, $5p$ (or $5d$), and $6s$ levels.

The relative isolation of the $4f$ electrons means that crystal field effects on the Nd^{3+} ion are of secondary importance. This secondary influence gives rise to small oscillator strengths and small splitting of the electronic levels.

The bands observed in absorption correspond to transitions from the ground term $4_{I_{9/2}}$ to any of the higher levels. The luminescence spectrum is then given by the transition from the $4_{F_{3/2}}$ term to the levels $4_{I_{13/2}}$ (emission at 1.3 μm), $4_{I_{11/2}}$ (emission at 1.06 μm), and $4_{I_{9/2}}$ (emission at 0.9 μm).

It is, however, the emission at 1.06 μm that is the most important. Figure 202 shows transitions between the individual levels in energy diagram.

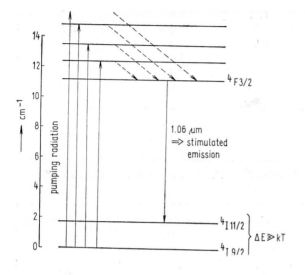

Fig. 202 Energy levels of the Nd^{3+} ion.

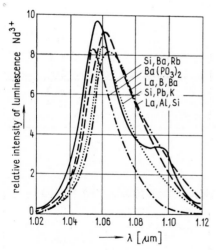

Fig. 203 Luminescent spectrum of Nd^{3+} glasses of various composition.

Measurement of luminescence is one of the basic prerequisites for the assessment of the quality of a glass laser. Figure 203 plots the relative values of the intensity of the Nd^{3+} luminescence for glasses of various composition [13].

Tables 52, 53, and 54 list further characteristic qualities of lasers made of glass of various kinds doped with Nd^{3+}.

Table 52

Components of glass	n_D	τ' [ms]	Luminescence [μm]	Width of the band [nm]
Si, Ba, Rb	1.529	0.81	1.057	26.0
Ba (PO$_3$)$_2$	1.593	0.26	1.054	24.5
La, B, Ba	1.656	0.08	1.061	36.5
Si, Pb, K	1.866	0.17	1.061	28.0
La, Si, Al	1.739	0.26	1.064	38.0
Ge, Ba, Rb, K	1.668	0.24	1.061	34.0

Table 53

Composition of glass: $(87 - x)$ mol. % SiO$_2$. xR$_2$O . 2.3 CaO . 6.5 BaO . 0.8 ZnO . 2.8 B$_2$O$_3$. 0.2 Sb$_2$O$_3$. 0.4 Nd$_2$O$_3$

R	x (mol. %)	τ' [ms]	Luminescence [μm]	Width of the band [nm]
Na	10	0.45	1.061	29.0
	19	0.49	1.059	28.5
K	10	0.62	1.059	29.0
	19	0.59	1.058	27.5
Rb	10	0.63	1.059	31.0
	19	0.66	1.058	28.0
Cs	10	0.62	1.059	32.0
	19	0.61	1.058	28.5

Table 54

Composition of glass: 69.5 wt. % SiO$_2$; 0.5 wt. % Nd$_2$O$_3$; X wt. % BaO; Y wt. % K$_2$O

X	Y	τ' [ms]	Luminescence [μm]	Width of the band [nm]
25	5	0.64	1.050	29.0
20	10	0.72	1.059	28.5
15	15	0.74	1.058	27.0
10	20	0.78	1.057	25.5
5	25	0.77	1.056	26.5

274

The lifetime, τ', is another characteristic property of glass lasers. It can (after [14, 45]) be defined with the help of the relationship

$$\frac{1}{\tau'} = P_{\text{rad.}} + P_{\text{non-rad.}}, \tag{302}$$

where $P_{\text{rad.}}$ denotes the probability of spontaneous transitions from the respective level per unit of time,

and $P_{\text{non-rad.}}$ the probability of non-radiative transitions from the respective level per unit of time.

Apart from the radiative transitions associated with emission, in Nd^{3+} there also exist non-radiative transitions representing losses in excitation energy. These can be expressed by the quantum yield

$$\eta = \frac{P_{\text{rad.}}}{P_{\text{rad.}} + P_{\text{non-rad.}}} = P_{\text{rad.}} \cdot \tau'. \tag{303}$$

Another parameter is the so-called threshold of stimulated emission. Stimulated emission occurs if the population of the $4_{F_{3/2}}$ level (denoted as N_2) is superior to that of the $4_{I11/2}$ level (denoted as N_1) in the ratio

$$\frac{N_2 - N_1}{V} \doteq \frac{N_2}{V} \geqq \frac{2\,\Delta v \tau'}{\lambda^2} \left(\frac{1-\varrho}{1} + \alpha\right), \tag{304}$$

where Δv is the linewidth of spontaneous emission,

λ the wavelength at which stimulated emission occurs,

ϱ the reflectance of the front surface of the resonator of length l,

V the volume of the resonator,

and α the absorption factor characterizing the internal losses in the resonator.

The threshold intensity, Φ_p, can be computed from relationship (305), if the pulse duration of the lamp $\Delta t > \tau'$

$$\Phi_p = h v_p \frac{\Delta N}{\tau' \eta}, \tag{305}$$

where $h v_p$ is the mean energy of the photons for the pumping radiation,

and τ' the lifetime of luminescence.

If we take account of the fact that from the intensity of the radiant flux, Φ_0, incident on the resonator only

$$\Phi = \Phi_0 \left[1 - \exp\left(-al\right)\right] \tag{306}$$

is absorbed, and if we disregard the losses due to reflection, we obtain for the incident threshold intensity of radiant flux the relationship

$$\Phi_p \approx \frac{2\,\Delta v h v_p}{\lambda^2} \frac{V}{\eta} \left(\frac{1-\varrho}{l} + \alpha\right) \frac{1}{1 - \exp\left(-al\right)}, \tag{307}$$

where a denotes the mean linear absorption coefficient.

275

The above equations hold for a four-level laser containing Nd^{3+} ions.

Glass lasers are frequently used as oscillators. They are very similar to crystal lasers and have a number of advantages, though their disadvantages can hardly be considered negligible.

Some advantages of glass lasers are the simplicity of production of rods, variation of optical parameters with glass composition, and a superior optical homogeneity of the active medium. "Giant pulses" require a particularly high degree of homogeneity.

The main disadvantage of glass lasers are the much larger and inhomogeneously broadened emission lines resulting in an inferior spectral purity of the radiation generated. The rather low thermal conductivity of glass sets limits to the maximum converted net power (viz. power output minus losses) and, consequently, to the maximum continuous power output. Despite much larger active volumes maximum pulse output is also lower than in crystals.

But even with such limitations glass lasers are capable of generating instantaneous power approaching that of ruby lasers.

6.8   CONDUCTORS OF RADIATION IN THE OPTICAL RANGE

In recent years glass fibres have increasingly been used for conducting optical radiation. The effect on which this application is based is total internal reflection. Use is generally made of [15]:
  (a) elementary glass fibres,
  (b) bundles of glass fibres,
  (c) multifibres, i.e., individual groups of clad fibres or rods,
  (d) bundles of fibres (even of double clad fibres) fused into a solid block.

In glass fibres the propagation of optical radiation is governed by the principle of total internal reflection, which occurs if the refractive index of the glass is higher than that of the surrounding medium (glass, air, resin, etc.).

Figure 204a is a schematic representation of the propagation of radiation through the conductor, disregarding, of course, refraction at the inlet and outlet surfaces.

The carrying capacity of the conductor depends upon the angle at which the radiation enters the conductor. The criterion is the so-called numerical aperture of the system, A'. In Figure 204b we present a schematic representation of the passage of radiation through the conductor with a surrounding glass layer.

When the ray of light meets the first surface at angle u_1, it is refracted:

$$n_1 \sin u_1 = n_c \sin u_c. \tag{308}$$

The ray then continues towards the interface between the glass and the surrounding medium with the refractive indices n_c and n_0, respectively, and is again refracted, viz.

$$n_c \sin \gamma_c = n_0 \sin \gamma_0. \tag{309}$$

For the critical angle γ_{ck} of the total internal reflection it holds that

$$\sin \gamma_0 = 1; \quad \sin \gamma_{ck} = \frac{n_0}{n_c}. \tag{310}$$

Considering the passage of the rays in $\triangle ABC$, we get

$$\cos u_{c.\,0} = \sin \gamma_{ck} = \frac{n_0}{n_c}, \tag{311}$$

or

$$\sin u_{c.\,0} = (1 - \cos^2 u_{c.\,0})^{1/2} = \frac{1}{n_c} (n_c^2 - n_0^2)^{1/2}, \tag{312}$$

where $u_{c,0}$ is the critical angle between the ray of light and the axis of the conductor.

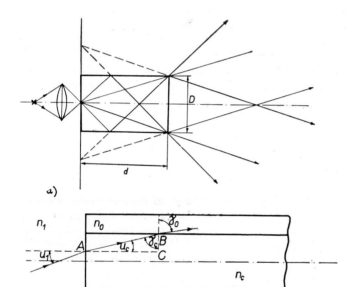

a)

b)

Fig. 204(a) Diagram of the passage of light through the conductor. Refraction in functional surfaces is not represented. (b) Diagram for the derivation of numerical aperture.

Then

$$A' = n_1 \sin u_{10} = n_c \sin u_{c.\,0} = (n_c^2 - n_0^2)^{1/2}, \tag{313}$$

where u_{10} is the critical angle for the refractive index n_1.

For conductors without an enveloping layer (the fibre being surrounded by air), the numerical aperture, A', can be calculated from the relationship

$$A' = (n_c^2 - 1)^{1/2}. \tag{314}$$

277

Thus, the higher the index of refraction of the glass conductor as compared with the refractive index of the enveloping layer, the higher the value of the numerical aperture. Table 55 [15] lists types of optical glasses, for conducting radiation.

Table 55

| Fibre | | Cladding | | Numerical aparture |
optical glass	n_d	optical glass	n_d	A'
BaF 7	1.608	BK 12	1.516	0.54
F 7	1.625	BK 12	1.516	0.58
SF 2	1.648	BK 12	1.516	0.65
LaK 11	1.658	BK 12	1.516	0.67
SF 5	1.673	BK 12	1.516	0.71
SF 4	1.755	BK 12	1.516	0.88
SF 4	1.755	FK 1	1.470	0.96

For transmitting radiation in the ultra-violet region of the spectrum use can be made of quartz glass fibres. For the infra-red region of the spectrum chalcogenide glasses, e.g., the As−S system, can be employed, and for the long-wave range of the infra-red region of the spectrum those of the As−Se−Te system.

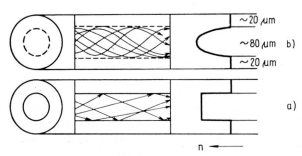

Fig. 205 Diagram visualizing the passage of light through a fibre with a step profile of the refractive index (a) and a gradient profile of the refractive index (b).

The problems of conducting optical radiation by glass fibres over long distances (so-called "optical communication") has been dealt with in outline by Gliemeroth, Krause, and Neuroth [33]; our presentation will therefore follow theirs.

As already mentioned above, fibres with a graded profile of the refractive index were originally used for conducting light, particularly high purity quartz surrounded by air or a lower index glass (see Figure 205a). However, the application of fibres in optical communication stimulated increased demand for a precisely defined gradient profile of the refractive index. This profile must satisfy a certain theoretically given function, as shown in Figure 205b (after [33]).

Although at first only transmission losses were considered, the development of the gradient profile fibres made it also necessary that transmission capacities should

278

be considered. The reduction of transmission losses to less than 8 dB/km has been achieved in fibres with a graded index of refraction. Table 56 (after [33]) lists the concentrations of impurities and their effect upon the losses due to absorption in glass produced by the MCVD method.

Table 56

Component	Absorption λ [nm]	Concentration of admixtures [p. p. m.] causing attenuation of approx. 10 dB/km at $\lambda = 800$ nm
Cu^{2+}	800	< 0.01
Cu^{+}	200	> 10
Fe^{3+}	300	> 10
Fe^{2+}	1100	> 0.03
Cr^{3+}	600	< 0.02
V^{5+}	720	0.008
M^{3+}	450	0.42
Co^{2+}	680	0.4
Ni^{2+}	410	0.35
OH^{-}	720	—
OH^{-}	820	30
OH^{-}	945	—

Apart from these impurities, and their effect upon the absorption, other losses must also be considered, viz. those due to light scatter, to bends of the conductors, and to geometrical variations of the section of the fibre.

The LED diodes (light-emitting diodes) are currently used as sources of light-emitting, non-coherent and weakly rectified radiation in the 750 nm — 900 nm waveband. Use is often also made of the so-called "semiconductor injection lasers" as coherent sources of light with an approximately equal range of wave-lengths.

The transmission capacity is determined by the stability of particular modes and their dispersion. The stability of the modes is inversely proportional, to the occurrence of imperfections in the fibre, and is thus dependent upon the homogeneity and continuity of the gradient of refractive index.

The dispersion depends on the geometric properties of the fibre, on the material applied, and on the difference between the indices of refraction of the cladding and the core of the fibre (viz. on the profile of the gradient of refractive index).

The transmission capacity is determined from the broadening of the signal after its passage through the fibre, i.e., by comparing the pulse time (half-width) before entry and after emergence from the fibre; it ought not to exceed 3 ns/km for transmission of 100 M bits per sec. over a distance of about one kilometer.

The sectional geometric asymmetry of the fibre should not exceed 1 % of the section.

In a homogeneous medium the pulse broadened is related to

$$\beta = n_c(\lambda) \frac{2\pi}{\lambda},$$
(315)

where λ is the wavelength of the light,

and n_c is the refractive index of the glass.

The signal modulated in the fibre propagates at group velocity V_g, for which we can write

$$V_g = \frac{d\omega}{d\beta} = -\frac{2\pi c}{\lambda^2} \frac{d\lambda}{d\beta} = -\frac{c}{\lambda^2} \frac{1}{\dfrac{d}{d\lambda}\left(\dfrac{n_c}{\lambda}\right)} = \frac{c}{n_c - \lambda\left(\dfrac{dn_c}{d\lambda}\right)},$$
(316)

where c is the velocity of light in vacuo.

One can therefore calculate the difference between the passage of rays along the axis $\Delta\tau_{u_c}$, and at critical angle $u_{c,0}$ for total reflection in the fibre of length l, viz.

$$\Delta\tau_{u_c} = \frac{l}{V_g \cos u_{c,0}} - \frac{l}{V_g} = \frac{l}{V_g} \frac{1 - \cos u_{c,0}}{\cos u_{c,0}} =$$

$$= \frac{l}{c}\left(n_c - \lambda \frac{dn_c}{d\lambda}\right) \frac{1 - \cos u_{c,0}}{\cos u_{c,0}},$$
(317)

and approximately

$$\Delta\tau_{u_c} \approx \frac{l}{2n_c}(A')^2,$$
(318)

where $A' = n_c \sin u_{c,0} \sim \sqrt{2n_c \cdot \Delta n}$ is the numerical aperture.

As already mentioned above, this expression was derived for a fibre with a graded index of refraction (see Figure 204), and it was assumed that all directions were "filled" with light of equal intensity. The higher the numerical aperture, the higher the intensity of radiation that the fibre is able to accept. The time difference in the passage of the radiation through the fibre is then termed "mode dispersion".

Since non-coherent light sources were made use of, a lowering of the numerical aperture and a reduction of scatter could not be achieved, so gradient optical material had to be applied. The latter makes it possible to compare the times of passage of light through different parts of the fibre and to establish an equivalent optical path to that for geometrical paths of different lengths.

Non-coherent light sources emit over a range of wavelengths of width $\Delta\lambda$. If we expand V_g about the λ_0 then

$$V_g\left(\lambda_0 \pm \frac{\Delta\lambda}{2}\right) = V_g(\lambda_0) \pm \frac{\Delta\lambda}{2} \frac{dV_g}{d\lambda_0} =$$

$$= V_g(\lambda_0) \pm c\lambda_0 \frac{\Delta\lambda}{2} \frac{\dfrac{d^2 n_c}{d\lambda_0^2}}{\left(n_c - \lambda_0 \dfrac{dn_c}{d\lambda_0}\right)^2} =$$

$$= V_g(\lambda_0) \pm \frac{\lambda_0 \, \Delta\lambda}{2c} V_g^2 \frac{d^2 n_c}{d\lambda_0^2} = V_g(\lambda_0) \pm \Delta V_g, \tag{319}$$

then for the total difference between the times $\Delta\tau$ of passage along the axis of the fibre of the slowest and the quickest component we can write

$$\Delta\tau = \frac{l}{(V_g - \Delta V_g)\cos u_{c.0}} - \frac{l}{V_g + \Delta V_g} \approx$$

$$\approx \frac{l}{V_g}\left[\left(1 + \frac{\Delta V_g}{V_g}\right)\frac{1}{\cos u_{c.0}} - \left(1 - \frac{\Delta V_g}{V_g}\right)\right] =$$

$$= \frac{l}{V_g}\left[\frac{1 - \cos u_{c.0}}{\cos u_{c.0}} + \frac{\Delta V_g}{V_g}\frac{1 + \cos u_{c.0}}{\cos u_{c.0}}\right] =$$

$$= \Delta\tau_{u_c} + \Delta\tau_M. \tag{320}$$

Table 57

Type of glass (core)	Suprasil	FK 51	BK 7	SF 16
n	1.452	1.482	1.510	1.631
$\dfrac{dn}{d\lambda}$ [μm^{-1}]	$-1.56 \cdot 10^{-2}$	$-1.19 \cdot 10^{-2}$	$-1.77 \cdot 10^{-2}$	$-3.25 \cdot 10^{-2}$
$\dfrac{d^2 n}{d\lambda^2}$ [μm^{-2}]	$2.97 \cdot 10^{-2}$	$2.83 \cdot 10^{-2}$	$3.67 \cdot 10^{-2}$	$9.88 \cdot 10^{-2}$
$\Delta\tau_{u_c}$ $\left[\dfrac{ns}{km}\right]$	75.3	76.7	78.4	85.1
$\Delta\tau_M$ $\left[\dfrac{ns}{km}\right]$	3.4	3.2	4.2	11.2

Mass (material) dispersion

$$\Delta\tau_M \approx 2l \frac{\Delta V_g}{V_g^2} = \frac{l}{c}\lambda_0 \, \Delta\lambda \frac{d^2 n_c}{d\lambda_0^2}, \tag{321}$$

is thus ascertainable for every ray (mode) from the spectral width of the light source and the dependence of refractive index upon the wavelength (also in this case mode dispersion has been limited by the profile of the index of refraction).

281

$\Delta\tau_M$ thus gives the upper limit of the transmission capacity of the fibre, and it lies within the interval of $3-12$ ns/km for LED diode with a spectral width of 40 nm (see Table 57).

The adequacy of the profile of the refractive index depends upon various parameters and is a complex function of the radius of the fibre.

From the relationship

$$n(r) = n_0 \left[1 - 2\Delta \left(\frac{r}{r_k} \right)^\alpha \right]^{1/2} \approx n_0 \left[1 - \Delta \left(\frac{r}{r_k} \right)^\alpha \right],$$ (322)

where $n(r_k) = n_M \approx n_0 (1 - \Delta)$,
we get for $\alpha \to \infty$ an asymptotic transition to a profile with graded refractive index.

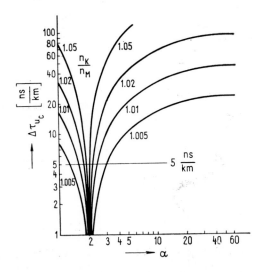

Fig. 206 Dependence of $\Delta\tau_{u_c}$ on α.

Figure 206 is a diagram of the mode dispersion, $\Delta\tau_{u_c}$, as a function of the exponent α; it follows that low values of $\Delta\tau_{u_c}$ can only be expected in the proximity of $\alpha \approx 2$. We can then write

$$|\Delta\tau_{u_c}|_{min} = \frac{n_0 \Delta^2}{8c} l,$$ (323)

for $\alpha = 2 - 2\Delta$.

For instance, for $\Delta\tau_{u_c} \leqq 5 \frac{\text{ns}}{\text{km}}$ and $A' = 0.26$, i.e. $\frac{n_k}{n_M} = 1.015$ to be reached it is essential that the value of α should be within the range

$$1.7 \leqq \alpha \leqq 2.3.$$ (324)

By the *transmission capacity of the fibres*, B, is understood the maximum transmissible sequence of binary signals per second over distance *l*. The resolvability of the binary information states 0 and 1 at the end of the fibre depends particularly on the type of code used, the form of the pulse, the type of excitation of the fibre, etc. A pulse length of 5 ns would thus correspond to a capacity of approximately 50 M bits/km/sec.

From equations (320) and (321) we get for the transmission capacity

$$B = \frac{1}{2\,\Delta\tau}\left(\text{or } \frac{1}{4\,\Delta\tau}\right) = \frac{1}{2(\Delta\tau_{u_c} + \Delta\tau_M)} \tag{325}$$

for a fibre with graded index from (318), and for the gradient fibre via exponent α from Figure 206. For the maximum possible velocity of bits we obtain for $\Delta\tau_{u_c} = 0$

$$B_{\max} = \frac{1}{2\,\Delta\tau_M} = \frac{c}{2l\lambda_0\,\Delta\lambda\,\dfrac{d^2 n_c}{d\lambda_0^2}}. \tag{326}$$

The problems of production of fibres for optical wave guides are discussed in detail in refs. [33, 34, 35, 36, 37, 38, and 39].

6.9   BALLOTINI

Glass ballotini are generally used in the production of reflex surfaces. Ballotini are glass pellets whose function is to focus rays of light emitted by a light source on to a point located in the vicinity of the rear surface of the pellet, viz. the surface at the interface between the pellet and the material on which the pellet is placed.

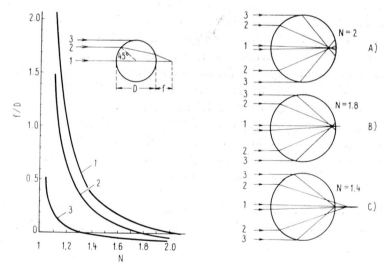

Fig. 207 Schematic diagram of the function of glass pellets (ballotini) as reflecting elements.

At that interface the rays get reflected. The basic requirement is for the pellets to be made of a glass having a high index of refraction. Figure 207 is a schematic diagram of the function of glass pellets as reflecting elements [17].

283

For a spherical lens, which the glass pellet virtually represents, we can express the focal distance (for axial rays) as

$$f = \frac{R(2 - N)}{(2N - 2)}, \qquad (327)$$

where f is the focal distance from the back surface of the pellet,

R is the radius of the pellet,

and $N = \dfrac{n}{n_0}$ (the effective index of refraction),

where n denotes the refractive index of the pellet,

and n_0 is the index of refraction of the environment in which the pellet is located.

For rays whose path through the pellet is distant from the axis the focal distance is shorter than the focal length for the axial rays. For practical purposes it suffices to know for which value of N the greater part of the luminous flux converges in the proximity of the back surface of the pellet. The calculation in Figure 207 has been made for three parallel rays and several values of N.

At $N = 2.0$ only axial rays converge on the back surface of the pellet; rays refracted at $45°$ converge on this surface at $N = 1.8$, marginal rays at $N = 1.4$.

At $N = 2.0$ all the rays, with the exception of a narrow band of axial rays, converge in the inner part of the pellet forming a circular area on the back surface.

At $N = 1.4$, on the other hand, all the rays, with the exception of the marginal rays, have their point of convergence behind the pellet, and at $N = 1.8$ all the rays are focused best on the back surface of the pellet. A closer study of this phenomenon will lead us to the conclusion that it is more important to attain a shorter focal distance for axial rays and for those refracted at an angle of $45°$, than it is for peripheral rays. It is therefore recommended that the (effective) index of refraction, N, should range between 1.8 and 2.0.

If the pellets are applied in such a way that only their rear parts are in contact with the light-reflecting layer, the front part of these pellets will lie in air, and refraction on the front surface will be defined solely by the axial index of refraction of the glass. For this case it is most advantageous to use glass whose refractive index $n = 1.9$.

If the pellets are located in a medium whose refractive index is higher than unity, it is recommended that use is made of glass whose refractive index exceeds 2.0.

References

[1] J. Nebřenský, Fotosenzitivní skla (Photosensitive glasses). *Informativní přehled SVÚS* No. 3, 1961.
[2] M. Fanderlik, Fotoplastická skla (Photoplastic glasses). *Informativní přehled SVÚS* No. 1, 1963.
[3] I. Fanderlik, Fotochromická skla (Photochromic glasses). *Informativní přehled SVÚS* No. 4, 1969.

[4] I. Fanderlik, The kinetics of colour centre formation and destruction in photochromic glasses containing silver halides. *Silikáty*, *14* (No. 3) (1970) 197—213.

[5] I. Fanderlik, Studie odmísení fáze halogenidů stříbra ve fotochromickém skle metodou elektronové mikroskopie (A study of the separation of silver halides in photochromic glass using electron microscopy). *Sklář a ker.*, *23* (No. 6) (1973) 165—170.

[6] V. V. Vargin, A. J. Kuznetzov, S. A. Stepanov, and V. A. Tzekhomskyi, Fotokhromnye stekla na osnove khlodistovo serebra (Photochromic glasses on the basis of silver halides). *Optiko-mekh. prom.*, *1* (1968) 35—72.

[7] F. P. Smith, Photochromic silver halide glasses. 7e Congres International du Verre, Bruxelles 1965, No. 108/III 2.

[8] J. Kocík, J. Nebřenský, and I. Fanderlik, Barvení skla (Colour Generation in Glass). SNTL, Prague, 1978.

[9] E. Lell, H. J. Kreidl, and J. R. Hensler, Radiation effects in quartz, silica and glasses. Progress in Ceramic Science, Vol. 4, Pergamon Press, London, 1966.

[10] N. M. Pavlushkin et al., Steklo spravochnik (Glass Handbook). Strojizdat, Moscow, 1973.

[11] I. Peyches, Les modifications spectrales entraínées par l'action des rayonnements sur les matériaux transparents. *Verres et Réfr.*, *22* (1968) 3, 261—267.

[12] J. Blabla, T. Šimeček, and V. Trkal, Kvantové generátory (Quantum Generators). SNTL, Prague, 1968.

[13] E. Snitzer, Glass lasers. Symposium on Coloured Glasses, Prague, 1967, 312—329.

[14] K. Pátek, Kvantové generátory světla a současný stav jejich výzkumu (Quantum generators of light. The present state of the art). *Pokroky matematiky, fyziky a astronomie*, IX (No. 4. (1965) 205—222.

[15] L. M. Kuchikyan, Svetovody (Wave Guides). Energia, Moscow 1973.

[16] J. M. Stevels, Neue Anwendungen des Glasses in Wissenschaft und Technik. *Wiss. Z.*, *23* (No. 2) (1974) 259—266.

[17] Swiss Patent No. 438600. Hochbrechendes Glass und dessen Verwendung. Corporation Searinght.

[18] I. Šolc, Silniční odrazová skla (Road traffic reflex glasses). *Sklář a ker.*, *10* (No. 11) (1960) 303—306.

[19] G. V. Byurganovskaya, V. V. Vavrgin, N. A. Leko, and N. F. Orlov, Deystvye izluchenya na neorganischeskye stekla (Effect of Radiation on Inorganic glasses). Atomizdat, Moscow, 1968.

[20] R. N. Dwivedi and P. Nath, Photochemical Reactions in Glasses Containing Silver. *Cent. Glass. Cer. Res. Inst. Bull.*, *24* (No. 3) (1977) 75—80.

[21] R. N. Dwivedi and P. Nath, Photochemical Reactions in Glasses Containing Gold. *Cent. Glass. Cer. Res. Inst. Bull.*, *25* (No. 1) (1978) 5—10.

[22] I. Fanderlik, Application of Greenwood's distribution and growth of isolated spherical particles theory on separated silver chloride phase in photochromic glass. (To be published in: *J. Non-Cryst. Solids*).

[23] S. Kumar and P. Ser, Optical absorption spectra of solarized Mn^{3+} and V^{2+} ions in glass. *Phys. Chem. Glasses*, *1* (No. 6) (1960) 175.

[24] S. D. Stookey, Coloration of glass by gold, silver and copper. *J. Am. Cer. Soc.*, *32* (No. 4) (1949) 246—249.

[25] S. D. Stookey and F. W. Schuler, Ultraviolet and X-ray irradiation effects in special photosensitive glass. IVe Congres International du Verre, Paris, 1956.

[26] M. H. Smithard and M. Q. Tram, Optical absorption produced by silver particles in glass. II Solid State Conference, Cairo, 1973.

[27] Polychromatic Glass — Corning Glass Works.

[28] P. J. Levy, Reactor and γ-ray induced coloring in crystalline quartz and Corning fused silica. *J. Chem. Phys.*, *23* (No. 4) (1955) 764—765.

[29] A. J. Cohen, Impurity-induced colour centres in fused silica. *J. Chem. Phys., 23* (No. 4) (1955) 765—766.

[30] C. Nelson and R. J. Weeks, Vacuum ultra-violet absorption studies of irradiated silica and quartz. *J. Appl. Phys., 32* (No. 5) (1961) 883—886.

[31] E. Lell, Radiation effects in doped fused silica. *Phys. Chem. Glasses, 3* (No. 3) (1962) 84—94.

[32] E. Lell, Synthesized impurity centres in fused silica. *J. Am. Cer. Soc., 43* (No. 8) (1960) 422—426.

[33] G. Gliemeroth, P. Krause, and N. Neuroth, Gläserne Telefondrähte. *Schott Information*, 1976, No. 2.

[34] Glass Fibre for Optical Communications — A selection of papers presented at the Society's Colloquium on Optical Fibres held in Imperial College, London 1979. In: *Phys. Chem. Glasses, 21* (No. 1) (1980) 1—66.

[35] K. Kobayashi, Optical and EPR studies on the interactions of Ag ions with polyvalent ions (As, Sb, Bi) and the formation of silver colloids. *Phys. Chem. Glasses, 14* (No. 1) (1973) 6—9.

[36] D. Küppers, J. Koenings, and H. Wilson, Codeposition of glassy silica and germania inside a tube by plasma-activated CVD. *J. Electrochem. Soc., 123* (No. 7) (1976) 1079—1083.

[37] A. Audsley and R. K. Bayliss, The induced plasma torch as a high temperature chemical reactor. I. Oxidation of silicon tetrachloride. *J. Appl. Chem., 19* (1969) 33—38.

[38] M. Achener and M. Habert, Fabrication en masse de silice pour fibre optique. *Fibre optiques, 56* (No. 12) (1976) 603—605.

[39] P. C. Schultz, Progress in the technology of wave guides and of the materials applied. Topical Meeting on Optical Fibre Communication, Washington, D. C., 1979.

[40] J. Mišek and L. Kratěna, Optoelektronika (Opto-electronics). SNTL, Prague, 1979.

[41] S. D. Stookey, G. H. Beall, and J. E. Pierson, Fullcolour photosensitive glass. *J. Appl. Phys., 49* (No. 10) (1978) 5114—5123.

[42] Ch. Rhee, Y. Jin et al., Studies on the optical properties of neodymium laser glasses. *J. Korean Phys. Soc., 11* (No. 2) (1978) 19—26.

[43] M. Tomozawa, R. H. Doremus, Treatise on Materials Science and Technology, Vol. 12, Glass I.: Interaction with Electromagnetic Radiation. Photochomic Glass (R. J. Aranjo). Academic Press, New York, San Francisco London, 1977, 91—122.

[44] M. Tomozawa, R. H. Doremus, Treatise on Materials Science and Technology, Vol. 17. Glass II.: Radiation Effects in Glass (E. J. Freibele, D. L. Griscom). Academia Press, New York, London, Toronto, Sydney, San Francisco 1979, 257—351.

[45] K. Pátek, In "Glass Lasers". Chemical Rubber Co., Cleveland—Ohio, 1970.

7 MODIFYING THE OPTICAL PROPERTIES OF GLASS BY THE APPLICATION OF LAYERS

7.1 INTRODUCTION

The rapid development in a number of industries has pushed up the requirements concerning the optical properties of glasses. This has particularly been the case as regards the application of glasses in the building industry, metallurgical industry, transport, optical instruments industry, etc.

Since it is not always fully possible, nor economically advantageous, to produce new glasses of special optical properties by a change in the chemical composition, the more usual way of attaining the requested optical properties has been the method of surfacing the currently produced glasses, viz. coating the glasses with transparent optical layers.

7.2 THEORY

Coating the surface of glass with transparent optical layers can substantially change the optical characteristics (see Chapter 4), the important aspects being the thickness, the properties, and the quality of application of the respective layer. In the following paragraphs we shall specify the terms "thin layer" and "thick layer".

7.2.1 Thin and thick layers

Using monochromatic radiation, we can, from the optical point of view, consider a layer of some tens of micrometers to be thin. The term "thin" can also be applied to a layer of much greater thickness, provided it is homogeneous. What matters is thus the quality of the applied layer: the coherent rays reflected at the interfaces of the layer should exhibit equal phase difference [1] for the whole area of the layer.

The definition of a thick and a thin layer becomes more accurate if the radiation is complex. In this case a layer is taken to be thin if the path difference of the rays of the medium region of the visible part of the spectrum traversing it does not exceed 5λ. This limit has been dictated by optical practice.

The difference in calculations on thin and thick layers lies in the fact that reflection on a thick layer is computed with the help of amplitudes, or intensities, only, whereas in thin layers not only the amplitudes but also the respective path, or phase, differences of the radiant rays are taken account of.

In thin layers interference effects can be observed, while in thick layers these effects do not occur at all.

From the relationship for the path difference

$$2n_1 l = 5\lambda, \tag{328}$$

where n_1 is the refractive index of the layer,
and λ is the wavelength of the radiation,

we can caculate the critical thickness of a thin layer for $n_1 = 1.5$ and $\lambda = 0.6 \ \mu m$

$$l = \frac{5\lambda}{2n_1} = 1 \ \mu m. \tag{329}$$

After Vašíček [1, 55] the thickness of a thin layer shall not exceed $l \leqq 1 \ \mu m$.

The reflectance of thick layers is then the mean value of the reflectance of the thin layer, as shown in Figure 208 (after [1]).

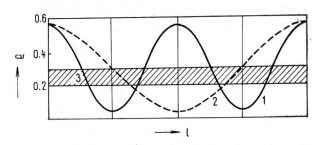

Fig. 208 Dependence of reflectance ϱ upon the thickness l of film layer for $\lambda = 400 \ nm \ (1)$, $\lambda = 800 \ nm \ (2)$, the mean value of the reflectance for thickly coated glass (3) $(n = 1.72 \ (glass)$, $n_1 = 1.35 \ (layer)$).

As already mentioned, the usual distinction is between *homogeneous layers*, where the refractive index remains constant through the whole thickness of the layer, and *non homogeneous layers*, where the refractive index undergoes variations dependent upon the thickness of the layers.

7.2.2 Interaction of radiation with glass coated with thick and thin layers

Thick layer

If the glass of index of radiation n is coated with a thick homogeneous layer the refractive index of which is n_1 and the radiation proceeding from the air medium ($n_0 = 1$) is perpendicularly incident upon this layer, then for the simple reflection, ϱ', at the air layer (viz. first surface) interface we can write (after [1])

$$\varrho' = \left(\frac{n_1 - 1}{n_1 + 1}\right)^2. \tag{330}$$

For the reflection at the second surface of the layer–glass interface, ϱ'', it holds analogously that

288

$$\varrho'' = \left(\frac{n - n_1}{n + n_1}\right)^2, \tag{331}$$

where n_0 denotes the index of refraction of the air ($n_0 = 1$),

n the index of refraction of the glass,

n_1 the index of refraction of the layer,

ϱ' reflectance at the air–layer interface,

and ϱ'' reflectance at the layer–glass interface.

We are now going to deal with the case of multiple reflection between the boundary surfaces of the layer. For simplification purposes, oblique incidence of the radiation on to a thick layer instead of normal incidence has been illustrated in Figure 209, because up to the angle of incidence $\Theta \doteq 40°$ reflectance does not differ very much from that for perpendicular incidence.

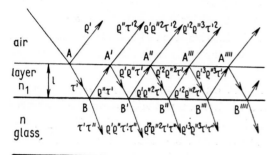

Fig. 209 Multiple reflection of radiation by a thickly coated glass.

The radiation meets the air–layer interface at point A. If we denote the reflection at the first refracting surface as ϱ' and the transmission as τ', then $\tau' = 1 - \varrho'$. The refracted ray then meets the layer-glass interface at point B, where it is split into two parts: the reflected part, $\tau'\varrho'$, and the refracted part, $\tau'\tau''$.

The ray refracted at point B returns to the interface point A', where it is again split into a reflected part, $\varrho'\varrho''\tau'$, and a refracted part, $\varrho''\tau'^2$, etc.

The radiation reflected from a thick layer is given by the sum of an infinite converging geometrical sequence ($\varrho'\varrho'' < 1$); thus

$$\varrho = \varrho' + \varrho''\tau'^2 + \varrho'\varrho''^2\tau'^2 + \varrho'^2\varrho''^3\tau'^2 + \ldots =$$

$$= \varrho' + \frac{\varrho''\tau'^2}{1 - \varrho'\varrho''} = \varrho' + \frac{\varrho''(1 - \varrho')^2}{1 - \varrho'\varrho''} = \frac{\varrho' + \varrho'' - 2\varrho'\varrho''}{1 - \varrho'\varrho''}, \tag{332}$$

because

$$\tau' = 1 - \varrho'. \tag{333}$$

And similarly, the radiation refracted by a thick layer is given by the sum of an infinite geometrical sequence

$$\tau = \tau'\tau'' + \varrho'\varrho''\tau'\tau'' + \varrho'^2\varrho''^2\tau'\tau'' + \varrho'^3\varrho''\tau'\tau'' + \ldots =$$

$$= \frac{\tau'\tau''}{1 - \varrho'\varrho''} = \frac{(1 - \varrho')(1 - \varrho'')}{1 - \varrho'\varrho''}, \tag{334}$$

because

$$\tau' = 1 - \varrho' : \tau'' = 1 - \varrho''. \tag{335}$$

The sum of the reflected and refracted radiation is

$$\varrho + \tau = 1. \tag{336}$$

Restricting ourselves to a single reflection, we can write

$$\varrho = \varrho' + \varrho''(1 - \varrho')^2, \tag{337}$$

and for the refracted radiation

$$\tau = \tau'\tau''. \tag{338}$$

Thin layer

Let us consider two parallel rays, S and S', incident on a thin layer of refractive index n_1 and thickness l coated on glass of refractive index n [1, 4, 5] (see Figure 210). The two rays travel in the same direction as the ray S'' and they interfere with one another.

The conditions for interference depend on the path difference between the two rays. Ray S' will cover the distance between point A' and point C, whereas the refracted ray S starting at point A will reach point F. The path difference between ray S and ray S' can be expressed as $\overline{FB} + \overline{BC}$. Since $\overline{BC} = \overline{BE}$, this path difference can be rewritten as the distance $\overline{FE}$. From the triangle CFE we get

$$\overline{FE} = 2l \cos \Theta_1, \tag{339}$$

where Θ_1, is the angle of refraction in the layer.

Since ray S passes through the layer of refractive index n_1, the optically equivalent path difference, Δ, between the two rays can be expressed as

$$\Delta = 2n_1 l \cos \Theta_1, \tag{340}$$

and for the respective phase difference δ we have

$$\delta = \frac{2\pi}{\lambda} 2n_1 l \cos \Theta_1. \tag{341}$$

If the path difference is an even multiple of half wavelengths, the reflection will be a maximum; if, on the other hand, the path difference is an odd multiple of half wavelengths, reflection will be a minimum. The relationship between the path difference and phase difference can be expressed as

$$\delta = \frac{2\pi}{\lambda} \Delta. \tag{342}$$

The interference between the two rays causes a change in the reflectivity and the transmittance of the system, which depend upon the refractive indices of the layer, n_1, and the glass, n. If $n_1 > n$, then the reflectivity of the system will increase relative to the reflectance of the uncoated glass. And, vice versa, if $n_1 < n$, the reflectivity will decrease (see Figure 211).

The effect of interference is in both these cases most pronounced if the optical thickness $n_1 l$ attains the value $\dfrac{\lambda}{4}$.

Let us assume that the radiation is perpendicularly incident upon the coated glass, that neither the glass nor the coating absorb the radiation, and that the surrounding medium (air) has refractive index $n_0 = 1$. For $\Theta = 0°$ the quoted relationships hold exactly but are only approximate for angles up to $\Theta \doteq 40°$.

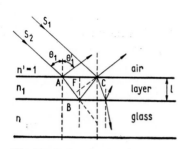

Fig. 210 Reflection of radiation by a thinly coated glass.

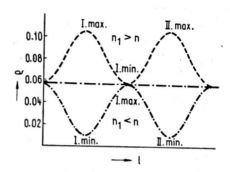

Fig. 211 Dependence of reflectance ϱ upon the thickness l of a thin coat for $n_1 > n$ and $n_1 < n$.

Applying the above simplifications we can (after [5]) write for the reflectivity R_1 on the first surface of the layer of thickness l_1

$$R_1 = \frac{\varrho_1^2 + \varrho_2^2 + 2\varrho_1\varrho_2 \cos \delta_1}{1 + \varrho_1^2\varrho_2^2 + 2\varrho_1\varrho_2 \cos \delta_1}, \tag{343}$$

where

$$\varrho_1 = \frac{n_1 - 1}{n_1 + 1}; \qquad \varrho_2 = \frac{n - n_1}{n + n_1}; \qquad \delta_1 = \frac{2\pi}{\lambda} 2n_1 l_1. \tag{344}$$

If the second surface is coated with a layer whose refractive index is n_2 and thickness l_2, the reflectivity R_2 can be written

$$R_2 = \frac{\varrho_3^2 + \varrho_4^2 + 2\varrho_3\varrho_4 \cos \delta_2}{1 + \varrho_3^2\varrho_4^2 + 2\varrho_3\varrho_4 \cos \delta_2}, \tag{345}$$

where

$$\varrho_3 = \frac{n_2 - n}{n_2 + n}; \qquad \varrho_4 = \frac{1 - n_2}{1 + n_2}; \qquad \delta_2 = \frac{2\pi}{\lambda} 2n_2 l_2. \tag{346}$$

The glass coated on both its surfaces has thus total reflection

$$R_{total} = \frac{R_1 + R_2 - 2R_1R_2}{1 - R_1R_2}. \tag{347}$$

7.3.1   Reflecting layers

The reflectivity of glasses can be increased by using metals, particularly silver, gold, aluminium, copper, rhodium, nickel, etc., and their alloys. With increasing thickness, however, the reflectance thus produced will reduce transmission. Layers the thickness of which exceeds 100 nm are usually not transparent. The reflectance ϱ of silver and gold layers are shown in Figures 212 and 213 as a function of thickness.

The spectral transmission of a Au layers has its characteristic maximum in the region of 500 nm [9]. This is why the human eye observes the transparency of a gold layer to have a greenish tint. In the infra-red region the curve of spectral transmission will decrease and the reflectivity curve will rise. In reflected light such a layer will have a golden tint (see Figure 213).

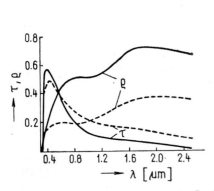

Fig. 212 Spectral transmission and reflection of the Ag (—) and Al (----) films. Insulating double pane.

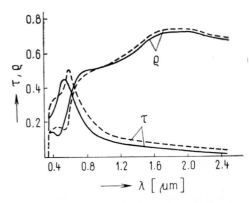

Fig. 213 Spectral transmission and reflection in an insulating double pane of two Float (6 mm) sheets of glass. The Au (—) and the Cu (----) coat on the inner side of the outer sheet.

A Cu layer (see Figure 213) shows a similar pattern of spectral transmission and reflection, the only difference being the fact that the maximum of transmission is placed nearer the longer wavelengths. A thin layer therefore has a yellowish tint in transmission and a cupric tint in reflection.

In the case of an Al layer (see Figure 212) we can observe a loss of selective reflectivity in the infra-red region of the spectrum.

Figure 214 shows the spectral characteristics of the following alloy: 75 wt. % Fe, 20 wt. % Cu, 5 wt. % Al [9].

The reflection of radiation from a thin layer of electrically conductive metal is accounted for by the presence of free charge carriers [9].

The classical Drude equation for the so-called *plasma frequency*, ω_p, can be used to give the boundary between the transmission and reflection of radiation. This frequency is dependent mainly upon the concentration of the free charge carrier, N; thus

$$\omega_p = \left(\frac{N e^2}{\varepsilon_0 \varepsilon_g m^*} - \gamma^2 \right)^{1/2}, \tag{348}$$

where e is the electronic charge,

       m^* the effective mass of the free charge carrier,

       ε_g the permittivity of the material without the free charge carrier,

       ε_0 the permittivity of free space,

       γ the damping constant $\left(\gamma = \dfrac{e}{\mu m^*} = \dfrac{1}{\tau} \right)$,

       μ the mobility of the free charge carrier,

and τ the mean period of impact.

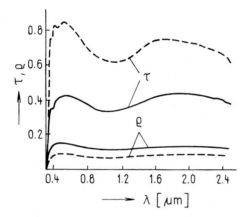

Fig. 214 Spectral transmission and reflection in an uncoated glass (---) and a glass coated with the alloy of 75 wt. % of Fe, 20 wt. % of Cu, and 5 wt. % of Al (—). Insulating double pane.

Since the effective mass of the free charge carrier, m^*, does not differ in practice from the mass of a free electron, the efficacy of the layer is given by both the concentration of the free charge carrier, N, and its mobility, μ. The higher the N and μ, the higher will be the electric conductivity and the reflectivity of the metallic layer. For the layer of the metal to manifest maximum reflection at the boundary between the visible and the infra-red region of the spectrum, it is essential that the concentration of the free charge carrier should be (after [9]) of the order of $2 \cdot 10^{22}$ [cm^{-3}]. On the other hand, the optical behaviour of metals in the ultra-violet and visible regions of the spectrum is affected by the absorption effects of the bonded electrons. The mobility of the free charge carrier depends upon the quality and the structure of the metallic layer applied. The disturbances of the structure (crystal lattice) are therefore accompanied by a restriction of the mobility, and, consequently, also by impairment of the reflective properties.

High values of the reflectance and, simultaneously, very low values of the absorption factor can be reached by a system of *dielectric layers* manifesting *selective reflectivity* in a certain range of the spectrum, dependent upon the material applied [7]. If an approximately 70 % reflection is to be achieved, it is however necessary to make use of substances with refractive index $n_1 \sim 4$; but the choice of these substances is restricted. Table 58 quotes the indices of refraction, the reflection maxima, and the values of the reflectance for one coat of the substances that have been applied on to the surface of glass.

Table 58

Substance	Refractive index n_D	Maximum reflection λ [nm]	Reflectance ϱ
Zn S	2.30	546	0.31
TiO_2	2.60	546	0.40
Sb_2O_3	2.70	1000	0.43
ZnSe	2.89	—	—
Ge	4.0	2000	0.69
Te	5.0	4000	0.79

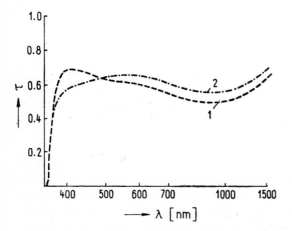

Fig. 215 Spectral transmission in the Reflex S (*2*) and Reflex KP (*1*) glasses.

The calculations of the reflectance of layers applied to the surface of glass are fully valid for monochromatic radiation only. For complex radiation we get selective reflectance, and interference colours will occur. The colour of the reflected and transmitted unabsorbed radiation is complementary. Composite, multilayer systems make it possible to increase reflectance up to the maximum, theoretically computed, values.

Reflecting glasses produced by the application of dielectric layers show higher chemical stability and mechanical strength as compared with metal-coated glasses.

It thus follows from the theory that the thickness of the layer can be adjusted and suitable substances applied, so that reflectance can to a certain measure be confined to the particular region of the spectrum desired.

In Figure 215 the reader will find the spectral characteristics of the Czechoslovak protective glasses coated with dielectric reflecting layers (titanium dioxide applied by dipping in a solution of organic Ti compound).

7.3.2   Antireflection layers

As already stated above, the reflective capacity of a glass-layer system will be decreased if the refractive index of the layer is lower than the refractive index of the glass, viz. if $n_1 < n$. The maximum reduction of reflectivity by a single layer can be achieved if the optical thickness of the layer equals $\dfrac{\lambda}{4}$ [7]. The substances most frequently used in the production of antireflection layers are magnesium fluoride, sodium aluminium fluoride, calcium fluoride, silicon dioxide, etc. The reflectance curve of a glass (of refractive index $n = 1.52$) coated with a layer of magnesium fluoride ($n_1 = 1.38$) is presented in Figure 216 [7].

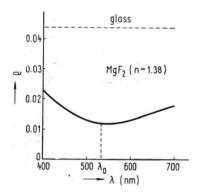

Fig. 216 Reflectance curve for an MgF₂ coated glass.

Owing to the unavailability of suitable substances, a zero minimum of reflectance for a certain wavelength, λ_0, can however not be achieved. Such a minimum can only be achieved for two, three, or more wavelengths by the application of a system of two, three, or more layers. The reflectivity over the whole spectrum can then be lowered below the level attainable by a single layer [7].

The properties of these layers [7] can, with some approximation, be studied with the help of a vector diagram. The vector sum of the amplitudes of the waves reflected at the interfaces then gives the resulting amplitude; zero reflectance is achieved if the vectors form a triangle.

Using three layers, we thus get three reflectance minima, the optical thicknesses of the layers in this case being $\dfrac{\lambda_0}{4}$, $\dfrac{\lambda_0}{2}$, $3\dfrac{\lambda_0}{4}$, respectively (from the upper layer). For the zero minimum, λ_0, there are two minima symmetrically placed round λ_0. Thus for a system of layers having optical thicknesses of 1, 2, ..., $n\frac{\lambda_0}{4}$ the reflectance curve is symmetrical with respect to λ_0 and it has n minima.

The reflectance curve of a triple layer of $MgF_2 - ZnS - Al_2O_3$ is presented in Figure 217 (after [7]).

The theoretical and practical problems of antireflection layers have generally been concerned with the applicability of the solutions to the glass elements of optical systems. For instance, attention has long been paid to the problem of facilitating the propagation of light through optical systems by reducing reflection.

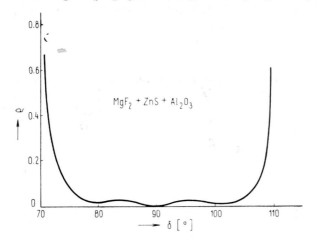

Fig. 217 Reflectance curve for a triple coat of MgF_2—ZnS—Al_2O_3.

A triple layer of magnesium fluoride − zinc sulphide − aluminium oxide is capable of reducing reflection over almost the whole spectrum below 0.05 %. A still more drastic reduction of reflection could be achieved by the application of further layers, but owing to reproducibiliy problems three has invariably been the highest number of layers applied (with the exception of the production of interference filters).

7.4   ABSORBING LAYERS

Absorbing layers applied to the surface of glass usually absorb in the visible range. These layers form, according to [10], the so-called *transient optical layers*, in which radiation is both absorbed and reflected. (As already mentioned above, radiation is, of course, also reflected by an uncoated surface of glass).

The absorption of the radiation is due to so-called colour centres and it depends upon the concentration of these centres and upon the thickness of the layer containing the colour centres [4]. The attenuation of the radiation traversing the absorbing layer can be calculated using the Lambert-Béer law (see Section 4.3.2) if the absorption due to the glass is disregarded or subtracted.

As absorbing layers are used the oxides of metals, e.g., nickel, copper, iron, chromium, cobalt, and other compounds, applied to the surface of the glass pure or as components of amorphous (vitreous) layers. The layers may be applied, for instance, by thermal decomposition of vapours or aerosols of the compounds of metals on to heated glass, by the gelling technique, by an electrochemical process, or by other methods used for upgrading utility glasses.

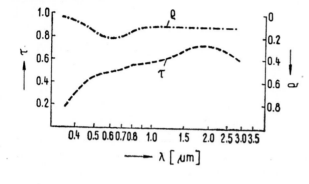

Fig. 218 Spectral transmission and reflection in a bronze grey Spectrofloat glass.

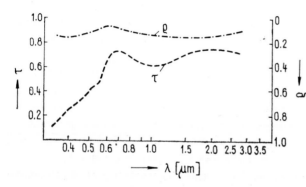

Fig. 219 Spectral transmission and reflection in a bronze cuprous Spectrofloat glass.

Figures 218 and 219 give spectral curves for the so-called Spectrofloat absorbing glass (electrochemically produced).

It is a disadvantage of the absorbing layers used particularly for architectural purposes that they absorb a certain part of the incident radiation, accumulate it, and re-emit it.

7.5   APPLYING OPTICAL LAYERS TO GLASS

The optical properties of a glass product depend to a high degree upon the quality of the applied layer, which is in turn dependent upon the method of the application of that layer.

The layer should have a uniformly arranged structure, it should exhibit equal thickness over the whole coated surface (except for cases of special decoration techniques), the refractive index should not vary with the thickness of the layer, the coating should adhere well to the surface of the glass, etc.

In view of such requirements each coating method has its particular field of application as far as the functional parameters of the coating layer, the technical difficulties and the quality of the application of the layer are concerned. We shall therefore briefly discuss some of the basic methods of applying layers on to the surface of glass, although we are fully aware of the fact that there exists a large number of other, modified methods.

7.5.1   Cathodic sputtering

By this method the substance that is to be used for coating the glass becomes the cathode of the glow discharge in an inert gas with pressures ranging between 13.3 and 1.3 Pa, and a voltage of several kV [2, 3]. The glass to be coated is placed on the anode. The positive ions arising from the discharge are accelerated towards the cathode, which they bombard at a velocity acquired in the region of the cathode drop. The magnitude of the normal cathode drop depends upon the type of the gas and the material of the cathode, and it grows non-linearly with the input voltage. The impact of an ion upon the glass to be coated is associated with an intense rise in the temperature and subsequent evaporation, although the mean temperature of the cathode may be comparatively low [2, 3].

The kinetics of cathodic sputtering may be explained by a pulse mechanism, according to which energy transferred by the ions has virtually the character of a shock wave propagating along the row of the atoms tightly arranged in an atomic grid. Some of the energy is transferred to the surface, and if it is larger than the bond energy of a surface atom, it is released. The quantity of metal sputtered is then inversely proportional to the pressure and the cathode-anode distance.

This method is particularly suitable for coating glass with metals and metallic alloys.

7.5.2   Ionic plasma sputtering

Here, the pressure ranges between 0.13 and $13 . 10^{-4}$ Pa. The auxiliary cathode is heated and the discharge path is in a magnetic field. The input voltage equals several hundreds of volts only [2, 3]. The glass specimen is not placed on the anode but on a special electrode, which is bombarded by the ions if its potential is negative. The velocity of sputter is relatively high, and the method yields uniformly deposited layers. The process of deposition can be controlled by means of the voltage on the sample. The method is generally used for applying layers of metals, alloys and compounds of aluminium oxide, cadmium sulphide, etc.

7.5.3 Vaporization in vacuo

Where the quality requirements, particularly those concerning the purity and structure of the applied layer are highest, use is made of the method of vaporization in vacuo [2, 3]. The atoms of the vaporized substance move rectilinearly, unless they collide; in order that the vaporized particles strike the sample, their mean free path must be longer than the distance between the source and the glass. A vacuum of $13 \cdot 10^{-6}$ Pa is normally applied.

The purity and the structure of the layer can be controlled via the pressure of the residual gases, the velocity of vaporization, the temperature, and the structure of the glass to be coated. This method, too, is used for applying coats of metals, metallic alloys, and other substances.

All the methods dealt with so far find application in the optical and the semi-optical industry, the imitation jewellery industry, and wherever high quality of the layers is the desired goal.

7.5.4 The gelling technique

This method is particularly suitable for applying dielectric layers. The esters of certain inorganic acids dissolved in a solvent convert by hydrolysis into gels, by drying into xerogels, and by burning into oxides of amorphous form [5]. The amorphous phase forms at temperatures much lower than the melting points of the respective substances. For instance, gelation of a solution of tetraethyl silicate and the subsequent dehydration give rise to silicon dioxide, viz.

$$Si(OR)_4 \rightarrow (OR)_3SiOSi(OR)_3 \rightarrow SiO(OR)_2n \rightarrow$$
$$\rightarrow SiO_2n\,H_2O \rightarrow SiO_2. \tag{349}$$

Since gelation can prevent considerable quantities of metallic salts from crystallizing before evaporating, this fact is taken advantage of in the application of silicon dioxide coats containing, for example, iron, trioxide, cobalt monoxide, nickel monoxide, molybdenum monoxide, vanadium pentoxide, etc., and titanium dioxide layers containing, for example, iron trioxide, copper monoxide, molybdenum monoxide, gold, platinum, palladium. The thickness of the layer depends upon the viscosity of the solution, the concentration of active components, the rate of withdrawing the glass out of solution (if this is the method used), and the speed of evaporation of the solution, etc. The gelling technique combined with the soaking technique (lifting the glass out of the solution) is a cheap way of applying optical layers of a large assortment, and it can particularly be used for producing protective reflecting glasses of large dimensions, e.g., for architectural purposes.

7.5.5   The electrochemical method

This is a very progressive method of coating thin optical films on to the surface of glass. It has been widely applied in the production of sheet glass by the Spectrofloat process. The method is based upon the electrochemical mechanism of diffusion of Cu^+ and Pb^{2+} ions from molten alloy into the surface of the glass [15, 16, 17, 18, 19, 20, 21]. The direct current makes the Cu^+ and Pb^{2+} ions diffuse from the alloy melt into the surface of the glass in exchange for Na^+ ions. The latter return to the tin bath where they are reduced, in the presence of H_2 in the protective atmosphere of N_2, to metallic Na. But the Cu^+ and Pb^{2+} ions are also reduced to Cu^0 and Pb^0 in the diffusion-applied layer forming larger aggregates and giving the glass a characteristic colouration.

The absorptive and reflective properties of a glass with a coating layer applied in the above way are dependent upon the thickness of the layer, the concentration and the dimensions of the metallic particles, and the nature of these particles. All these parameters can be controlled via the temperature, time, and the concentration of H_2 in the protective atmosphere of N_2.

Use can also be made of a Ni/Bi alloy supported by a nickel anode, a Ag/Bi alloy, etc.

The so-called Chameleon technique is also based upon the electrochemically governed diffusion of the colouring metals into the surface of the float glass, the only difference being the multicoloured ornaments, which have given the technique its name [19, 20, 21].

7.5.6   Applying layers by spraying

The surface of glass can also be sprayed. The glass can be pre-heated, in which case the vapours, or aerosols, of the metallic compounds undergo decomposition [14].

The method of spraying the layer on to the surface of the glass is frequently applied in coating sheet glass; it is, however, also made use of in upgrading and finishing utility glass products, etc.

7.5.7   Chemical application of layers

The chemical application of layers is based upon the reduction of metals (e.g., gold, silver, and others) from solutions with the help of a suitable reducing agent [18]. This method is similar to that of making mirrors, where the silver layer is produced, for example, from an ammoniacal solutoin of silver nitrate and sodium hydroxide, and the reducing agent can be, for example, tartaric acid, lactose, Rochelle salt, etc. The silver layer is then applied from the mixed solutions on to a clean surface of the glass. The respective chemical reaction begins after mixing of the two solutions, with the production of silver hydroxide. The drop

300

in the number of Ag^+ ions gives rise to decomposition of the silver complex, and ammonia is released.

The silver layer is then given a protecting coat of a suitable varnish, or is galvanically coated with a film of copper. The process of silvering can be accelerated by elevated temperature or by the use of rapidly acting reducing agents.

7.5.8   Ionic exchange

This method, originally developed for strengthening glasses by means of a surface layer, can also be used for applying optical coatings. The exchange of ions in the surface of the glass is a process of diffusion obeying Fick's laws. The number of the ions exchanged at constant temperature is directly proportional to the square root of the time of thermal treatment, and increases with the temperature [23].

The ionic exchange is invariably brought about by immersing the glass in a bath of fused salt, mixed salts, or by applying a mixture of plastic clay, and the diffusing substance on the surface of the glass.

For instance, exchange of the Na^+ ions in the glass for Ag^+ ions from the melt $(AgNO_3, AgNO_3 + KNO_3,$ etc.) proceeds in such a way that the Ag^+ ions penetrate into the surface of the glass, are reduced owing to the presence of certain components in the glass, and form aggregates of $(Ag^+)_n$ to Ag^0. Depending upon the temperature of the bath and the duration of immersion, the glass coating acquires a colour ranging from yellow via brown to red. For instance, in the process of cuprous staining the alkalis are exchanged for copper (used in the form of cuprous chloride). The number of substances that can be introduced into the surface of the glass by means of ionic exchange is fairly large.

When the problem of reinforcing glasses using ionic exchange was being solved, attention was also paid to the spectral characteristics of the silver layers applied to the surface of Float glass by exchanging Na^+ ions for Ag^+ ions from a bath of 2 wt. % of silver nitrate and 98 wt. % of potassium nitrate. Depending upon the temperature of the bath and the duration of the immersion, layers were obtained of thickness 10 µm − 51.5 µm and exhibiting characteristic spectral transmission. For particular results the reader is referred to Figure 220.

Glasses with a photochromic layer of silver halides have recently been also produced using the method described above [24]. In order that glasses melted without silver (but containing chlorine, bromide, fluorine, and copper monoxide) may have photochromic properties (viz. formation of microcrystals of silver halides), they are immersed in a bath of $AgNO_3$, $AgNO_3 + NaNO_3$, or $AgNO_3 + AgCl$ at a temperature of 400 °C (for aproximately four hours). After the ionic exchange of $Na^+ \rightleftarrows Ag^+$ has been completed, the glass is thermally treated, first at a temperature of 400 °C (for eight hours) and then at 660 °C (for forty minutes). The photochromic layer formed by the microcrystals of the silver halides is 100 μm – 200 μm thick. When exposed to radiation (of the sun or of an artificial source), these glasses darken and discolour. The degree of darkening is proportional to the intensity of the radiation.

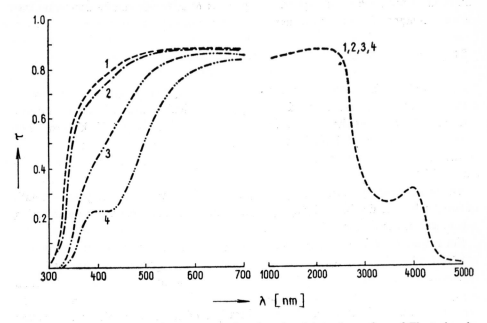

Fig. 220 Spectral transmission in a coat of silver introduced into the surface of Float glass by ionic exchange: *1* — temperature of the bath: 470 °C, duration 2.5 min; *2* — temperature of the bath: 470 °C, duration 5 min; *3* — temperature of the bath: 470 °C, duration 20 min; *4* — temperature of the bath: 470 °C, duration 60 min.

Another way of coating glasses with a silver halide layer is first to apply the layer of silver from a solution with a complex-forming component and reducing agent present, and then to halogenate with a solution of copper chloride or copper bromide.

Layered photochromic glasses can also be produced by applying organic foils (or by inserting the foil between two sheets of glass), which become progressively more coloured as the intensity of radiation increases.

The disadvantage of practically all photochromic organic materials is the fact that they are fatigue-prone: they manifest fatigue after cycling through several hundred of reversible photochromic processes.

7.5.9 Burning layers into glass surfaces

Some methods of applying metallic layers and their oxides to the surface of glass are based on the diffusion of the ions, or they are at least accompanied by processes of diffusion [25, 26]. One of these methods, by which the metals are

302

burned into the surface of the glass, or the diffusion of the ions into the surface of the glass is at least accelerated by elevated temperature, is particularly used for giving a final finish to utility glass products.

In the steady state diffusion obeys Fick's First law, and in non-stationary states Fick's Second law. The concentration gradient, which is the driving force of the diffusion, is however considered for ideal systems only, or for systems approximating to the ideal system. For other systems the gradient of chemical potentials must be used. In multiphase systems, diffusion tends to equalize the chemical potentials of the diffusing substance in the individual phases; it does not however equalize the concentrations in the whole system. The assumption of a constant diffusion coefficient throughout the whole mass is generally not valid. In practice, Fick's Second law must be employed in a modified form [25].

The diffusion coefficient thus characterizes the diffusivity of a substance. The diffusion of the ions into the surface of the glass depends upon their dimensions, the structure of the surface, the type of diffusing substance, and upon external conditions. The metallic ions diffused into glass penetrate to a certain distance, are reduced to neutral atoms due to the effect of the medium, cluster into nuclei, and grow into colloidal particles, which then generate characteristic colouration. This method is extensively used for giving a finish to utility glass and other glass products.

7.5.10 Silk screen printing

The principle of the silk screen process consists in rubbing a paste through the meshes of a stencil placed on a screen stretched in a frame. A screening tool puts the stencil into contact with the surface to be printed upon. The pressure of this tool and its uniform sliding motion across the surface of the stencil rolls the paste forward squeezing it through the meshes of the screen. The screening tool is made of rubber or some other resilient material and the angle of its inclination to the plane of the stencil determines the quantity of paste that reaches the surface.

The thickness of the paste which has been passed through the screen, measured after being dried and fired, depends not only upon the thrust of the screening tool, but also upon the thickness of the silk screen, the thickness of the stencil layer, and upon the proportion of volatile solvents in the paste.

The silk screen printing method is mostly used for marking a decorating container and utility glass products. It must, however, be emphasized that layers applied in this way can in no way be considered entirely homogeneous from an optical point of view, and their utilization therefore depends upon the requirements concerning the quality of application.

7.5.11 Deposition of layers from gaseous phase

One of the methods of applying layers from the gaseous phase is the vaporization method made use of mainly for finishing utility glass, imitation jewellery and other glass products. The vaporizing substance, mostly $SnCl_2$, $TiCl_4$, $Sr(NO_3)_2$, $BaCl_2$, $Bi(NO_3)_3$, $Cu(NO_3)_2$, is vaporized (in the form of vapour mist) on the pre-heated product at ambient atmospheric pressure, or in vacuo. Evaporation takes place in special ovens, chambers, or drums, and various colours can be generated depending upon the thickness of the layer applied. The layers of uniform thickness are homochromatic, and those of varying thickness heterochromatic. If a layer of uniform thickness is to be achieved, vaporization must be carried out in vacuo. The thickness, thus also the hue, depend upon the composition of the vaporizing mixture, the temperature and the duration of the process.

The vaporization method is used for giving a decorative finish to utility glass products and imitation jewellery.

The advantageous properties of tin dioxide layers can be exploited in finishing container glass. Apart from giving the product irridescent colouration, the tin dioxide layer reinforces it against internal overpressure, and the products manifest improved "slipping" contact.

Use can, for example, be made of the $(CH_3)_2 SnCl$ compound, which decomposes to SnO_2 in the vaporizer. It is fed into the vaporizer by means of a current of hot air at $60° - 140 °C$. The equipment employed for this type of coating glasses can produce homogeneously deposited layers.

7.5.12 Etching the surface of glass with acids

The antireflection layer can in this case be produced by leaching alkalis out of the surface of the glass, mostly with chloroplumbic acid (H_2PbCl_4). The thickness of the layer and, consequently, the colour of the coating layer are dependent on the temperature and the duration of the leaching. Thus, for example, leaching the BaF6 optical glass in a 75 °C H_2PbCl_4 bath will generate the following colours, which correspond to different periods of leaching:

53 min	yellow
63 min	bluish violet
82 min	green
100 min	iridescent

Layers produced by this method are of particular interest for the optical industry.

**VARIOUS APPLICATIONS OF THIN LAYERS
IN THE OPTICAL INDUSTRY**

   The right combination of several reflecting and antireflection layers
applied to a surface of colourless or coloured glasses can produce interference
filters, which are used as simple monochromators separating a narrow spectral
region from a complex luminous flux [1].

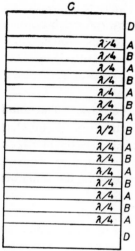

Fig. 221 Schematic diagram
of a 15-layer interference
filter.

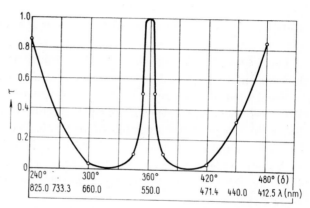

Fig. 222 Dependence of the transmission in a 15-layer
interference filter upon the phase difference δ and the
wavelength λ of the radiation. If $\delta = \dfrac{2\pi}{\lambda}\, \Delta$, then for 550 nm
(if $\Delta = \lambda$) $\delta^\circ = 2\pi = 360^\circ$. Secondary maxima can be
absorbed with the help of adequate filters.

   Figure 221 gives a schematic diagram of an interference filter consisting of
fifteen layers.

   The central layer, $\dfrac{\lambda}{2}$, with path difference $\Delta = \lambda$, has a low refractive index
(layer B). Its thickness $l = \dfrac{\lambda}{2n_2}$ determines the wavelength for which the filter
is maximally transmissive. The central layer is adjoined by layers $\dfrac{\lambda}{4}$ with path
difference $\Delta = \dfrac{\lambda}{2}$ manifesting a high or low index of refraction according to
the sequence $A, B, A, \dots B, A$. The thickness of layers $A(l_1)$ are given by equation
$l_1 = \dfrac{\lambda}{4n_1}$, those of layers $B(l_2)$ by the relationship $l_2 = \dfrac{\lambda}{4n_2}$.
   The refractive index of the glass that supports the layers is n_0. Figure 222 shows

305

the dependence of the spectral transmission of this interference filter upon the phase difference (or the wavelength).

Use is also made of interference filters with frustrated total reflection [1, 7]. Frustrated total reflection appears if a low refractive index layer is placed between two optical prisms and if the angle of incidence of the radiation exceeds the critical value (with a thick layer we would get unimpaired total reflection). If the layer is thin ($l < \lambda$), total reflection is impaired and part of the radiation continues its passage towards the other prism. With the right choice of the thickness of the layer, the radiation can be split into two parts, the reflected part and the part passing in a certain direction; and if we disregard the absorption, this separation is free from any losses, Thus, for example, using glass prisms with $n_0 = 1.72$ and a layer of MgF_2 ($n_2 = 1.36$), the critical angle of incidence will equal 53.2°, and a 60° prism can be used. But multilayer systems, e.g., $MgF_2 - ZnS - MgF_2$ ($n_{ZnS} = 2.3$) have, of course, a much wider field of application.

Polarization of radiation can also be attained with thicker layers. Thin films, however, make it possible to produce large scale polarizers, the function of which is based upon the principle of plan-parallel plates located in a beam of rays at Brewster's angle of incidence. The right selection of the layer and of its refractive index can suppress the reflection of one polarized component of the radiation, at the same time increasing the reflection of the other.

A simple polarizer can, for example, be produced by applying titanium dioxide layers to either side of the glass, in which case the degree of polarization will equal 0.99.

Separating polarization filters are based on the principle of splitting a beam of light without losses due to absorption by a system of dielectric layers applied between two glass prisms. The right choice of index of refraction of the two prisms, using, e.g., ZnS-cryolite layers, can ensure a degree of polarization of the reflected light higher than 0.99.

In optical practice, a parabolic surface was often the desired form of non-spherical surface; it was obtained by a modification of the shape of the closely related spherical surface. Thin film layers were used for parabolizing mirrors. The shortcoming of these parabolizing layers, e.g., aluminium films, was the fact that at thicknesses exceeding 2 μm radiation was scattered. This can be avoided by using layers based on the principle of Fresnell's mirrors, each layer being applied in vacuo by means of a rotating diaphragm.

Of still greater importance is the method of producing off-axis paraboloids, which are made use of wherever refracting elements are to be eliminated (the infra-red region of the spectrum).

Layers have also been used for correcting aberrations of lenses, with the upper limit on thickness of the layer, e.g., an LiF layer, being about 5 μm, because of diffusion effects. If, however, colloidal layers are applied, this limit can be exceeded as much as four times.

The diffraction of radiation sets limits to the sharpness of the definition of optical images. The diffracted image consists of a central and a secondary maximum. The secondary maxima can be suppressed by the insertion of a screen made up of thin layers.

The method of vaporizing antireflection layers can be applied in correcting the curves of spectral transmission of the optical systems, sun glasses and other protective glasses. In some cases the requirement is for thin layers used as modifiers to exhibit a precisely defined transmittance. A mathematical solution of the problem is rather complex and, moreover, one that imposes a number of limitations. For instance, pre-determined indices of refraction of the individual layers can be employed and the required optical properties of the system of thin layers attained by calculating the thickness of these layers, or vice versa.

The ophthalmic glasses industry makes use of reflecting and absorbing layers in the production of non-dioptric or dioptric protective sun glasses. The layers are invariably applied to the surface of the glass by vacuum techniques.

7.7 MEASUREMENT OF THICKNESS AND PROPERTIES OF OPTICAL LAYERS

7.7.1 Measurement of thickness

The thickness of a layer determines the spectral characteristics of the coated glass, so that it is advisable that the thickness of the layer should continually be checked.

The simplest way of measuring the thickness of a layer is to observe the interference colours in white light. The thickness is then calculated from the relationship

$$l = \frac{m\lambda}{2\sqrt{n_1^2 - \sin^2 \Theta}},$$
(350)

where m denotes the order of interference,
λ the wavelength (relative),
n_1 the refractive index of the layer,
Θ the angle of incidence of the light
and l the thickness of the layer.

Another way of determining the thickness of a layer is by weighing. If the area of the applied layer and its specific mass are given, the thickness of the layer applied can be calculated from the difference between the masses of the glass before and after the application, viz.

$$l = \frac{\Delta M}{\varrho P},$$
(351)

where l denotes the thickness of the layer,

$\quad\quad\Delta M$ the difference between the masses of the glass before and after the application of the layer,

$\quad\quad\varrho$ the specific mass of the layer,

and $\quad P$ the area of the applied layer.

The thickness of a layer can be measured by taking advantage of the effect of the changing frequency of a piezo-electric crystal as a function of the thickness of the layer. The frequency of the crystal forming a substrate for the layer varies with the thickness of the layer being applied.

The most precise method of measuring the thickness of layers is Tolansky's method. It utilizes the interference effects between the layer to be measured and a glass cover.

Eddy currents can also be utilized in measuring the thickness of thin layers. If the metallic layer is placed in an alternating magnetic field, eddy currents are induced the intensity of which is proportional to the total resistance of the layer; for a given metal the resistance of a layer depends upon the thickness of the layer [41].

A very simple method of measuring the thickness of silver layers consists in placing an iodine crystal on the surface of the glass freshly coated with silver [40]. After some time the iodine crystal is surrounded by rings. A certain number of these rings corresponds to a certain thickness of the silver layer.

From among many other methods we should mention those making use of radioactive isotopes and β-radiation (the Betascope apparatus).

The X-ray method of measuring the thickness of thin layers is classified under non-destructive methods. The reduction in intensity of the X-ray radiation is proportional to the thickness of the film being applied (and, of course, to the absorption of the X-ray radiation by the substance applied).

For a description of these methods see ref. [23]. The thickness of a layer and its index of refraction can also be measured by ellipsometers.

7.7.2. Measurement of adhesion to glass

The measurement of the adhesion of the optical layers applied to the surface of glass makes use of a number of methods depending upon the function for which the coated glass has been designed.

The simplest of these is the method of determining adhesion qualitatively by sticking a particular adhesive tape to the layer and tearing it with the layer from the glass surface.

The adhesion of an aluminium layer is measured by a method applying a small ball of gold pressed into the layer; the force expended in order to separate that ball from the layer is then measured.

The method of drawing a spike across the layer to be measured is the most universal. The spike is curved and is made of steel or other material. It is drawn across the layer and gradually loaded until the layer gets separated from the glass [42]. The weight of the load is a measure of the adhesion.

The loading can also be kept constant, in which case the measurement consists in counting the number of journeys of the spike across the layer needed to separate it from the glass, or in recording visually, with the help of a microscope (or another optical, e.g., interference, method), the depth of the notch depending upon the number of the journeys that the spike has made.

Another method is the measurement of resistance to abrasion. The resistance of the layer to abrasion is recorded for various materials under standard conditions. The method can also be applied in combination with the effects of certain chemical solvent, dust etc.

7.7.3 Measurement of optical properties

The spectral transmittance of glasses coated with transparent optical layers can be measured in any region of the spectrum by almost all types of modern spectrometers.

Spectral reflectivity may be measured by integrating spheres. In some cases spectrophotometers can also be adapted for this purpose.

The transmission and reflection of solar radiation can be measured by both calorimetric and thermoelectric methods [44].

In the case of the *calorimetric method*, use is made of direct solar radiation as the source. Two calorimeters are used: the one at the side facing the source (solar radiation) glazed with a three-milimeter thick reference glass of known transmittance, the other housing the glass to be measured. When an equilibrium has been established in the two meters, the heat is removed by a circulating current of air the flow and the temperature of which are recorded. From the recorded values thermal losses due to the glazing and the insulation of the calorimeters are then computed. The screening factor, S_c, is defined by the ratio of the magnitude of the loading of the measured glass by solar heat to the solar load of the reference glass (3 mm thick) under the same contitions of irradiation. If, for the 3-mm-thick reference glass, transmission $\tau_j = 0.87$ has been recorded, then

$$S_c = \frac{\tau_{ef}}{0.87} 100 \quad [\%]. \tag{352}$$

From the recorded values of S_c effective transmission, τ_{ef}, can be calculated.

The method of *thermoelectric measurement* is based on the principle of recording solar radiation passed through, or reflected by, the glass under examination. The instrument is controlled by a clock mechanism mounted paralactically with respect to solar radiation to ensure perpendicular incidence upon the glass sample. The radiation that has passed through, or has been reflected, is recorded by a thermo-element the sensor of which has an absolutely black surface; the sensitivity of the sensor is uniform over the whole range of the solar spectrum. A sensitive galvano-meter and the thermoelement form a closed system. Figure 223 is a schematic diagram of the apparatus.

An iris diaphragm (*1*) defines a parallel beam (*B*) of solar radiation, which is aimed, by means of a tube (*3*) with a shutter (*2*) at the sample of glass (*4*) placed in a swivel support (*5*). The radiation that has passed through is recorded by thermoelement (*6*), reflected radiation is recorded in position (*6′*). The radiation incident on the thermoelement before the measurement has started is denoted by Φ_0; after the sample has been placed in the measuring position the transmitted and the reflected radiation are denoted by Φ_t and Φ_r, respectively.

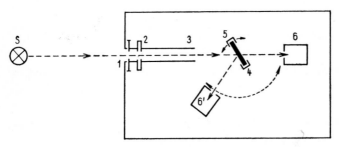

Fig. 223 Schematic diagram of an apparatus measuring transmission and reflection of solar radiation in glass (thermoelectric method).

The angle of incidence of solar radiation can be adjusted by the console support-ing the sample up to an angle of 80° for measurement of transmission and to an angle of 10°−70° for measurement of reflection. The values recorded are then used for calculating transmittance, reflectance, and absorptance.

$$\tau = \frac{\Phi_t}{\Phi_0}, \tag{353}$$

$$\varrho = \frac{\Phi_r}{\Phi_0}, \tag{354}$$

$$1 = \tau + \varrho + \alpha. \tag{355}$$

The values τ, ϱ, and α are primary. In order to calculate (after [47]) the effective transmission of solar radiation, τ_{ef}, standard conditions must be considered, viz. natural convection on the internal side of the glazing, $\alpha_1 = 8.2$ W/m² . deg, forced convection on the external side of the glass given by a velocity of 3.3 m/sec, $\alpha_e = 22.5$ W/m² . deg.

References

[1] A. Vašíček, Optika tenkých vrstev (The Optics of Thin Layers). NČAV, Prague, 1956.

[2] L. Eckertová, Tenké vrstvy. Fyzikální základy a některé aplikace ve fyzice (Thin layers. Physical foundations and some applications in physics). Jemná mech. opt., 11 (No. 8) (1966) 214—218.

[3] L. Eckertová, Fyzika tenkých vrstev (The Physics of Thin Layers). SNTL, Prague, 1973.

[4] I. Fanderlik, Optické vlastnosti skel (The Optical Properties of Glasses). SNTL, Prague, 1979.

[5] M. Kříž, Tenké vrstvy na skle vytvářené gelovou technikou (Thin layers applied on glass by the gelling technique). Sklář. keram., 22 (No. 7) (1972) 169—172.

[6] B. Havelka, Geometrická optika (Optical Geometry). I, II, NČAV, Prague, 1956.

[7] Vl. Vrba, Tenké vrstvy v optice (Thin Layers in Optics). Jemná mech. opt., 3 (No. 12) (1958) 409—416.

[8] S. Minář, Některé problémy technologie protiodrazových vrstev (Some problems of the technology of antireflection layers). Jemná mech. opt., 3 (No. 6) (1958) 188—191.

[9] R. Groth, Vakuum beschichtete Sonnenschutzgläser für Bauindustrie. Glastechn. Ber., 50 (No. 10) (1977) 239—247.

[10] M. Kříž, Možnost zlepšení ochranných vlastností skla proti sluneční radiaci (Possibilities of improving the protective properties of glass against solar radiation). 1st. Conference on Glass in the Building Industry, 23—24 Oct. 1973, Ústí on Elbe.

[11] Z. Knitl, Geometrie vakuového napařování (Geometry of vacuum steaming). Jemná mech. opt., 3 (No. 2) (1958) 62—66.

[12] S. Furuchi, Properties of solar energy reflecting films on glass composed of titanium dioxide and metals. Research laboratory, Asahi Glass Co., Yokohama.

[13] A. Thelen, Theoretical studies on multilayer antireflection coatings on glass. J. Opt. Soc. Am., 50 (No. 5) (1960) 509.

[14] K. T. Bondarev et al., Stekla dla zashchiti ot solnechnoy radiatsyi vypuskaynye za rubezhom (Protective Glasses against Solar Radiation Produced Abroad). V. N. nauchno-tekhn. inf. i ekon. prom. stroy. met., Moscow, 1973.

[15] M. Pakulska and K. Regula, Modyfikovanije powirzchni szkla w metodzie float (Modifying surfaces of glass by float method), Szklo Ceram., 26 (No. 4) (1975) 103—106.

[16] Pat. USA 3 472 641. Ford Motor Co. (R. D. Gray), Process for Ornamenting Glass Manufactured by the Float Glass Process.

[17] Pat. USA 3 582 402. Nippon Sheat Glass Co. (K. Hideo and M. Yoshiaki), Process and Apparatus for Continuous Manufacture of Reinforced Glass Ribbon.

[18] Collective of Authors, Tvarování plochého skla (Shaping of sheet glass). SNTL, Prague, 1974.

[19] Pat. GDR 104 498. Pilkington (W. Haltman and C. Howard), Verfahren und Vorrichtung zur Herstellung von Glas mit modifizierten Oberflächen, insbesondere kontinuierliche Herstellung von Flachglas sowie ein danach hergestelltes Glas.

[20] Pat. USA 3 850.605. Pilkington (W. Haltman and C. Howard), Manufacture of Patterned Glass.

[21] Pat. G B 1 382 837. Pilkington (W. Haltman and C. Howard), Patterned Glass.

[22] O. V. Vorobeva, Teplozashchitnye stekla s oksidnometallicheskimi pokrytyiami (Thermoprotective glasses with coatings of metallic oxides). Steklo VNIIS, (No. 1) (1964) 39—43.

[23] Vl. Novotný, Zpevňování skla (Reinforcing glass). SNTL, Prague, 1972.

[24] H. M. Garfinkel, Photochromic glass by silver ion exchange. Applied Optics, 7 (No. 5) (1968) 789—794.

[25] Z. Cozl and R. Zrůstek, Využití povrchové difúze iontů kovů pro barvení skla (Using surface diffusion of metallic ions for colour generation in glass). Symposium on Coloured Glass, Prague, 13th—16th September, 1967.

[26] J. Matoušek, Metody studia difúze iontů ve skle (Methods of studying ionic diffusion in glass). *Technický zpravodaj PUS, 11* (No. 3) (1969) 22—24.

[27] W. Klein, Sonnenreflexionsglas für den Hochbau. *Glas, 53* (No. 3) (1976) 91—94.

[28] P. M. Zavreckij and G. H. Foby, Transparent heat-mirror films of titanium dioxide (silver) for solar energy collection and radiation insulation. *Appl. Phys. Lett., 25* (No. 12) (1974) 693—695.

[29] Collective of Authors, Elektronika tenkých vrstev (Electronics of Thin Films). SNTL, Prague, 1970.

[30] J. Buřil, Přehled vakuových napařovacích přístrojů n. p. Laboratorní přístroje (Survey of the Vacuum steaming apparatus of Laboratorní Přístroje. Prague). *Jemná mech. opt., 11* (No. 8) (1966) 232—236.

[31] V. Vrba, Monochromatické interferenční filtry (Monochromatic interference filters). *Jemná mech. optika, 3* (No. 3) (1958) 75.

[32] V. Boček, Korekce křivky spektrální propustnosti optických soustav vhodnou volbou antireflexních vrstev (Correction of the curve of spectral transmission by a suitable choice of antireflection layers). *Jemná mech. opt., 3* (No. 3) (1958) 76—77.

[33] S. Minář, Vytváření vrstev kovových kysličníků na sklo (Applying coatings of metallic oxides to glass). *Jemná mech. opt., 11* (No. 1) (1966) 11—12.

[34] J. Schilder, K otázkam zloženia a štruktúry tenkých vrstev (Problems of composition and structure of thin films). *Jemná mech. opt., 11* (No. 4) (1966) 108—111.

[35] J. Schilder, K otázkam zloženia a štruktúry tenkých vrstev (Problems of composition and structure of thin films). *Jemná mech. opt., 11* (No. 6) (1966) 166—170.

[36] J. Červenák, O metóde katódového naprašovania (The method of cathodic sputtering). *Jemná mech. opt., 11* (No. 6) (1966) 164—166.

[37] Z. Knitl, Tenké vrstvy v optice (Thin films in optics). *Jemná mech. opt., 11* (No. 8) (1966) 218—220.

[38] S. Minář, Nové poznatky v technologii výroby vnějších odrazových vrstev na skle (New information on the technology of the manufacture of reflecting glass coatings). *Jemná mech. opt., 8* (No. 11) (1963) 353—357.

[39] Neues Schichtdicken-Messgerät. *Metalloberfleche, 17* (No. 8) (1963) 249—250.

[40] V. A. Ryabov and I. I. Borisova, Opredelenye tolshchiny serebryannykh zerkalnykh pokrytiy metodom ionnykh koles (Determination of the thickness of silver mirror coatings by the method of ionic rings). *Steklo VNIIS, 17* (No. 2) (1960) 23—25.

[41] S. M. Rabinovitch. Pribor dlya opredeleniya toschchiny serebryannovo sloya na zerkalakh (Apparatus for the determination of the thickness of silver layers on mirrors). *Steklo i keram., 19* (No. 7) (1962) 18—21.

[42] P. Benjamin and C. Weaver, Measurement of adhesion of thin films. *Proc. Roy. Soc.,* (No. 254) (1959) 163—176.

[43] V. Prtržílka, Fyzikální optika (Physical Optics). Přírodověd. nakl., Prague, 1952.

[44] M. Dubček, Přímé měření průchodu sluneční energie sklem (Direct measurement of the propagation of solar energy through glass). 1st Conference on Glass in Building Industry, 23—24 October 1973, Ústí on Elbe.

[45] F. Jelínek, Ploché sklo v obvodovém plášti budov (Sheet Glass in the Peripheral Envelopment of Buildings). SNTL, Prague, 1975.

[46] J. Puškáš, Tepelno-technické hodnotenie zasklenia klimatizovaných budov (Thermal Evaluation of the Glazing of Air-Conditioned Buildings). 1st Conference on Glass in Building Industry, 23—24 October, Ústí on Elbe.

[47] P. Hesoun, Tepelné zisky oslněnými okny (Thermal Gains by Glazed Windows). ČVTS-Committee for the Environment OS-1 Prague, 1971.

[48] S. K. Gupta and S. N. Prasad, Production of mirror and silvering on glass. Part II — Silvering methods. *Cent. Glass Cer. Res. Inst. Bull., 14,* (No. 1) (1967) 18—22.

312

[49] S. K. Gupta and S. N. Prasad, Production of mirror and silvering on glass. Part III — Coopering, painting and testing methods. *Cent. Glass Res. Inst. Bull., 14* (No. 2) (1967) 122—126.

[50] H. J. Becker and S. Schiller, Ochranná skla proti slunečnímu záření pro stavebnictví nanášením tenkých vrstev na ploché sklo ve vysokém vakuu (Protective glasses against solar radiation for the building industry produced by the application of thin films on sheet glass in deep vacuo). 1st Conference on Glass in the Building Industry, 23—24 October 1973, Ústí on Elbe.

[51] V. Vrba, Antireflexní fázová vrstva (Antireflection phase layer). *Jemná mech. opt., 13* (No. 3) (1968) 91—94.

[52] L. Vašková and S. Kvapil, Povrchová úprava svářečských skel vrstvami zlata (Coating of welder's glasses with layers of gold). *Sklář. a keram., 23* (No. 12) (1973) 364—366.

[53] V. Gottardi and A. Raccanelli, Analysis of thin films on glass by nuclear techniques. *Glass Technology, 17* (No. 1) (1976) 26—30.

[54] ČSN Czechoslovak Standard, Světelně technické názvosloví (Lighting terminology). 1967.

[55] A. Vašíček, Optics of thin Films. North Holland Publ. Co., Amsterdam, 1960.

[56] Z. Knitl, Optics of Thin Films. J. Willey a Sons, SNTL, London, New York, Sydney, Totonto, Prague 1976.

[57] G. Hass, R. E. Thun, Physics of Thin Films. Acad. Press. New York, London 1967.

[58] O. S. Heavens, Optical Properties of Thin solid Films. Butterworths, London 1955.

[59] I. Fanderlik, R. Obršálová, Transparentní optické vrstvy na sklech (Optical Transparent coatings on the Glass). *Informativní přehled SVÚS.* 1978, No. 4.

[60] I. Fanderlik, Transparentní optické vrstvy na sklech (Optical Transparent coatings on the Glass). 1st Konference "Povrchové zušlechťování skla", Jablonec n. N., 1979, 23—55.

Complex light, 35
Components, trichromatic, 38—39
Concentration,
    of fluctuation, 188, 189
    of OH groups, 162, 163
Configuration of electrons, 24, 25
Conductivity, electrical, 66
Conductor of radiation, 276
Constant,
    Boltzmann, 35, 188
    photoelastic, 218
    Planck, 35
    Stefan—Boltzmann, 34
Cooling, stabilization, 90
Coordinates of chromaticity, 41
Coordination number, 94
Couple, thermoelectric, 173
Coupling, Russel—Saunder's, 135
Critical angle, 68, 69, 277
Cryostat, 174
Crystal, rate of growth, 52, 53, 54, 55
Crystallization, 54
Curve of dispersion, 81, 82, 83

Decibel, 147, 279
Decomposition,
    by nucleation and growth, 56, 57, 58, 59
    kinetic theory of, 60
    spinodal, 55, 56, 57, 58, 59
Definition,
    Kirchhoff's, 34
    of absorption, 146
    of reflection, 68
    of refraction, 82
    of scatter, 194
    Stefan's, 34
Density, specific mass, 144
Depolarization, 197
Deposition of gaseous phase, 304
Detector of radiation, 172
Diagram,
    Orgel's, 137, 138, 139, 140
    uniform chromaticity scale, 41
Dichroism, 218
Diffraction grating, 170
Diode, light emitting, 279
Dipole moment, 79
Discriminating ellipse, 41
Dispersion,
    angular (grating), 171

anomalous, 81
curve, 81, 82, 83
equation, 80
mass, 281
mean, 84
mode, 279
partial, 85
relative, 85
Dissymmetry, 196
Double refraction, 217

Effect,
    Bezold—Brücke's, 37
    Jahn—Teller's, 141
    of polarization, 212
    Purkyně's, 37
    Tyndall's, 186
Electrochemical method, 300
Electron,
    configuration, 24, 25
    energy, 16, 20
    frequency, 26, 79
    mass, 78
    optical polarizability of, 79
    transition of, 25
    velocity of, 19, 78
    vibration of, 78, 79
Ellipse, discriminating, 41
Emission,
    spectra, 35
    spontaneous, 28
    stimulated, 28, 272
Energy,
    interaction, 129
    inter-phase, 53, 54
    light-quantum, 33, 34
    of activation, 53, 54
    of crystallization, 53, 54
    of electron, 129
    of excitation, 275
    of rotation, 129
    of vibration, 129
    potential, 18
Enthalpy, free, 52
Entropy,
    molecular, 54
    of crystallization, 52
Equation,
    of dispersion, 80
    Schrödinger, 17, 18, 19, 21

315

polarizability of electron, 92, 93
properties of layers, 309
Oscillator,
    classical, 21
    harmonical, 20, 81
    strength, 26
Ostwaldian process of maturation, 60
Oxygen,
    bridging, 95
    non-bridging, 95

Parameter, Racah's, 135
Partial dispersion, 85
Partial vapour pressure, 160
Paschen's notation, 25
Path,
    actual length of, 218
    difference, 218, 221
    optical, 218
Permeability, 72
Permittivity, 66
Phase,
    difference, 191
    separation
        metastable, 55
        stable, 55
    velocity, 78
Phosphorescence, 270
Photoluminescence, 270
Polarization, 212
    Brewster's angle of, 215
    by absorption, 218
    by reflection, 214
    by refraction, 216
    by scattering, 216
    circular, 213
    elliptical, 213
    of radiation, 212
    plane, 213
    rotary, 219
Polarimeter, 218
Polariscope, 218
Polarizability, 86, 87, 93
    optical, 79
    static, 79
Point,
    melting, 50
    softening, 112, 115
Pressure, partial water vapour, 160
Principle of correspondence, 22

Prism,
    angular dispersion of, 170
    characteristic dispersion of, 170
    dispersion of light by, 170
    Pekhanov's, 114
Printing, silk screen, 303
Properties of layer (optical), 307
Purity, colorimetric, 40
Purkyně's effect, 37

Quantity, luminous, 48
Quantum number,
    angular momentum, 16
    azimuthal, 16
    magnetic, 16
    principal, 16
    radial, 16
    spin, 16

Radiance, 49
Radiation,
    infra-red, 30
    interference, 220
    monochromatic, 35
    quantum theory of, 32
    ultra-violet, 30
    visible, 30
    wave theory of, 31
Radius,
    critical, 60
    ionic, 97
Range of transformation, 51, 85
Ratio, Rayleigh's, 197
Ray,
    reflected, 64
    refracted, 78
    transmitted, 143
Reaction, photochemical, 33
Reflectance,
    of aluminium, 67
    of gold, 67
    of silver, 67
    upon the absorption, 72
    upon the angle, 68
    upon the refractive index, 71
    upon the wavelength, 70
Reflection,
    angle, 68
    coefficient, 65
    diffuse, 67

319